Aquaculture project formulation

by
David Insull
Senior Fishery Planning Officer
FAO Fishery Policy and Planning Division
and
Colin E. Nash
Programme Leader
UNDP/FAO Aquaculture Development
and Coordination Programme

FAO FISHERIES TECHNICAL PAPER
316

Reprinted by arrangement with the
Food and Agriculture Organization of the United Nations

by

DAYA PUBLISHING HOUSE
DELHI-110035

FOOD AND AGRICULTURE ORGANIZATION OF THE UNITED NATIONS
Rome, 1990

"This book was originally published by the Food and Agriculture Organization of the United Nations (FAO) as Aquaculture project formulation, FAO Fisheries Technical Paper 316.

First Indian Edition 1995
ISBN: 978-81-7035-150-4

Rs. 340

Published by : **Daya Publishing House**
1123/74 Deva Ram Park,
Tri Nagar, Delhi-110035
Phone : 7220593, 7231826

Printed at : Raj Offset Works, Delhi

PREPARATION OF DOCUMENT

This document was prepared as part of a series of Technical Papers dealing with planning concepts and methodologies in the fisheries sector. Intended mainly as didactic materials for use in training courses, it provides an introduction to procedures and problems involved in the formulation of projects within the context of the special characteristics of the aquaculture sub-sector.

Please note the present address of Dr C.E. Nash is: P.O. Box 4606, Rolling Bay, WA. 98061, USA.

Distribution:
FAO Fisheries Department
FAO Regional Officers
Directors of Fisheries
Aquaculture (general)
Authors

Insull, D.; Nash, C.E.
Aquaculture project formulation.
FAO Fisheries Technical Paper. No. 316.
Rome, FAO. 1990. 129p.

ABSTRACT

This document is intended primarily as didactic material for use in training courses in aquaculture project formulation. It can also be read as a text in its own right by government administrators and planners, particularly in developing countries, and by commercial investors in aquaculture.

The first part of the document contains a broad introduction to project formulation, describing the integration of aquaculture projects within development plans, the organization and management of project formulation projects, and the stages of the project cycle. It is not only important for projects to be satisfactorily integrated into the economy of the sub-sector, but also that those responsible for project formulation should be aware of the practical problems which may arise in project implementation. Project formulation and implementation, therefore, are described briefly here as a single entity, consisting of twelve phases, and the more frequently occurring problems encountered in project implementation are described. Illustrations of three actual aquaculture projects are given to show the diversity which may be encountered by planners.

The second part of the document is concerned only with the six phases of project formulation, encompasing project identification, preparation, and appraisal. The sequence of activities carried out within each phase are described within 17 steps, each including further activities or tasks. Drawing on the characteristics of these illustrative models particular attention is given to differences of approach between the public and private sectors.

PREFACE

Aquaculture projects are similar in many respects to agricultural projects. They require similar formulation, the same standards of design and analysis, and experienced and trained personnel to carry out the formulation. As in agriculture, aquaculture projects are linked directly to specific sites, either on land or water, and are a form, directly or indirectly depending on the type of project, of rural development; they may similarly be adversely affected by pollution or by local economic or social changes which inhibit economic activity at their fixed location.

Nevertheless, aquaculture has many characteristics and criteria for development which are special to it. Moreover, those persons in developing countries who have responsibility for aquaculture develoment generally have a professional background in capture fisheries and an academic training in biology or another natural science; where officials have had training in aquaculture it has usually been in aspects of production rather than planning. They are thus unlikely to have contact with, let alone be familiar with, the procedures of aquaculture project formulation.

The purpose of this publication is to introduce those charged with aquaculture development to the underlying rationale of the project approach to overcoming development problems within the context of the aquaculture sub-sector. It provides step-by-step guidance to the process of project formulation, focussing on the relationships of the steps to each other, rather than explaining in detail the many techniques required in project formulation. For those wishing to study the subject more deeply, a list of publications for further reading will be found at the end of the text. A limited number of publications are cited here. Taken together, however, they describe most of the techniques used in aquaculture project formulation and appraisal.

A major purpose of this publication is to provide documentation for use in training courses dealing with capture fisheries and aquaculture project formulation and in courses concerned with aquaculture sub-sector planning and management. Its treatment of project formulation is therefore extended, in one direction to discuss the relationship between sector planning and projects, and in the other to describe briefly the function of project implementation and its relationship to project formulation. The latter is only a part of the project approach to development. If the project is not well integrated into the sectoral economy, a disappointing outcome is likely regardless of how competently the techniques of the formulation process have been applied. It is also essential that project planning takes the full account of practical realities of project implementation. The project formulation team, therefore, must be aware of the requirements of project implementation and potential areas of project weakness.

While a major concern of officials charged with aquaculture development will be with projects which are at least partially funded by official external assistance, the major part of investment in aquaculture is made by the private sector. Such investment is frequently dependent on previous government investment in infrastructure and almost certainly on institutional measures in support of the sub-sector. The linkage between private sector investment and government intervention and support means that officials should be aware of the key differences between public and private sector procedures in project formulation.

The publication is in two parts. In Part I, "An Introduction to Projects", provides an overview of the project approach. Its first chapter describes some aspects of projects within the development process and the ways different types of organizations might undertake project formulation and implementation. Chapters 2 and 3 describe the project formulation process and project implementation. The final chapter of Part I illustrates through case studies different types of projects within the aquaculture sub-sector.

Part II is divided into chapters which deal with the three main stages of the project cycle, i.e., Identification, Preparation, and Appraisal. Within these stages six phases are defined, each phase containing one or more steps. In all, 17 steps are described in the form of a series of tasks typical of that part of project formulation. The differences between the procedures in the public and private sectors are described at each step. The practical application of each phase is illustrated by reference to three imaginary projects, one dealing with a "green field" private-sector project, one with a large brackishwater-development project funded through external assistance, and one with a regional project providing assistance in institution building, improved production and enhanced product quality.

The advantage of such a step-by-step structure is that it provides a comprehensive view of the formulation process coupled with an insight into the use of the process in the rational and systematic collection of information, its analysis and the types of judgement which are required of the formulation team. A possible weakness is that it may suggest a mechanical process where there is no need or opportunity for creativity. Equally, the step-by-step explanation of the formulation process may convey the impression that a rigid sequence must be adhered to, work on each step chronologically following the previous one. In practice, neither of these conditions exist in project formulation . It is hoped, therefore, that the text contains sufficient emphasis on the need for an open-minded, innovative approach, particularly in the identification of projects, and that adequate attention is adequately drawn to occasions where certain steps may well be undertaken concurrently.

ACKNOWLEDGEMENTS

The basis of this publication is the FAO "Guide for Training in the Formulation of Agricultural and Rural Investment Projects", published in 1986, from which the methodological framework has been taken with some slight adaptation. The "Guide" has also been borrowed from closely where its treatment of a topic is particularly applicable to aquaculture project formulation. It is listed in the "Further Reading", at the end of this document; it is especially useful in its incorporation within one publication of descriptions of most of the techniques used in project formulation.

The need for good resource allocation and scheduling are emphasized in two documents which, together, have been the source of the chapter which deals with project implementation. These are "Health Project Management: a manual of procedures for formulating and implementing health projects", published in 1974 by the World Health Organization (WHO) which describes aspects of project implementation in different contexts, and a document published in 1979 by the US Agency for International Development (USAID), "Agricultural Project Implementation". This publication combines well documented examples and imaginary project situations, and presents a choice of options, an analysis of approaches, and the consequences of decisions. The importance of good design and, in particular, proper scheduling is also emphasized by another publication of which extensive use has been made. This is "The Design of Agricultural Projects: lessons from experience", produced by the FAO Investment Centre in 1989 as Technical Paper No. 6.

In describing some of the technical issues in aquaculture project design, much reliance has been placed on a document produced by the UNDP/FAO Aquaculture Development and Coordination Programme (ADCP), "Planning An Aquaculture Facility: Guidelines for bioprogramming and design", published in 1988.

A document produced by the Economic Development Institute (EDI) of the World Bank in 1984, "Aspects of Project Appraisal", was used as the basis of the chapter dealing with this subject.

Dr Michel Girin of France Aquaculture, the aquaculture development project arm of the Institut Français de Recherche pour l'Exploitation de la Mer (IFREMER), was responsible for preparing a working document which elaborated the structure adapted here, and also much of the textual materials, including a number of the project illustrations.

The authors also wish to thank those FAO staff members who have read and commented on drafts for parts of the publication.

CONTENTS

List of Tables

List of Figures

PART I. AN INTRODUCTION TO PROJECTS

1. PROJECTS WITHIN THE DEVELOPMENT PROCESS

A project is a scheme to organize the use of a given quantity of resources in a specific way to achieve particular results, all within a definite time. It has a precise beginning and a precise end. The execution of a project requires multidisciplinary effort, mobilizing different skills and resources to achieve predetermined development objectives which will result, directly or indirectly, in new or added value or social, economic or financial benefits.

Within this general definition many different examples of projects can be described, each with a different type of organization. However, they all have the same fundamental parts or phases, from the time each project is first conceptualized until it is terminated. These parts fit together and relate to each other in what is called the project cycle.

The Relationship Between Projects and Development Plans

Projects may arise from many sources. They may originate or be included within a sector study and from there be translated into the sector development plan; they may result from a programming mission carried out by an external aid agency, a follow-up to another project, a project idea originated in a government agency or an external assistance agency, a commercial initiative, etc.

Sector and project planners, however, should be constantly aware that, whatever the source, projects must be well-integrated into the planning process or difficulties may occur. Among potential difficulties are the following:

- There is no machinery for identifying and resolving critical policy issues.

- If projects do not respond to real needs, natural resources will be dissipated.

- Donor-financed projects may be accepted because of their short-term advantages to hard-pressed departmental budgets, but in the long-term there are insufficient resources to sustain them.

- There is a conflict with other activities when projects are implemented.

- Project objectives may conflict with each other.

Many of the weaknesses in projects which materialize in project implementation may be attributed to poor project formulation, in particular the often less-than-systematic way in which project ideas are integrated into the economic and institutional fabric of the sector. Good project formulation thus starts at the sectoral planning stage. The key to good sectoral planning is the sector study.

The sector study, into which national and sectoral objectives are fed, consists of two parts: (i) the sector review, or "stocktaking" of the sector, and (ii) the sector analysis, or "diagnosis". The sector review is a description of the sector presented in such a way that the analysis follows from it. The analysis identifies areas of development concern evident from the review, and then indicates opportunities, constraints, and policy issues which have to be resolved. One result of the sector study may be that some or all objectives for the sector, set at the national level, have to be revised. Thus one of the outputs of the study is quantification of sector objectives into real and attainable targets.

Targets have to be translated into plans. Methodology for this is widely accepted, although the terminology is often confusing. The word "strategy", for example, is often used in place of "policy", and vice-versa; and sometimes planners confuse the substance of strategies with that of policies, or promote them to being objectives in their own right.

The strategy is the key planning document. It may be appropriate for a country's aquaculture sector to have a single strategy. Alternatively, if there are a number of distinct and significant parts of the sector (such as marine shrimp and seaweeds), or there are different geographic or climatic zones (such as upland areas high in rainfall and dry coastal flatlands), then each may require differing approaches and development initiatives, and a separate strategy may be needed for each sub-sector. This may be exemplified by the following illustration.

> Country X has a national objective to improve the supply of fish from its inland fisheries. This is to be achieved through a single strategy encompassing both aquaculture and inland fisheries development. The strategy consists of three policies, each directed toward achieving certain specific aims, namely: (i) enhancement of culture-based fisheries, (ii) promotion of small-scale entrepreneurial fish-farming, and (iii) improvement of the capture fishery. These policies are delivered through policy instruments, which are (a) legislative changes, (b) provision of support services, and (c) projects. The policy concerned with the enhancement of culture-based fisheries consists of two types of policy instruments, namely projects and government support measures. Projects include construction of hatcheries and stocking of specified inland waters, improvement of landing sites, and provision of credit; government support services include improved management programmes and extension services targeted primarily toward small-scale farmers. Other policies, consisting of selected policy instruments, are formulated for small-scale fish-farming and inland capture fisheries.

In this illustration, the objective of improving the supply of fish from inland fisheries is first translated into a series of targets (production from inland fisheries, small-scale fish-farming, and inland capture fisheries), and these are the ends. The means by which the ends are met are: (i) the strategy, (ii) the policies, and (iii) the policy instruments.

It is important to note also that public sector projects are just one part of the development process. Equally, and possibly more important, is the total range of measures in addition to public sector investment projects which government has at its disposal to bring about profitable private investment.

Sectoral strategies, policies, and policy instruments are therefore all major components of the sector plan. The plan is concerned with the inputs required to obtain targeted outputs, and the management of these inputs. Thus, for the above policy concerned with small-scale fish-farming the plan might consist of a statement of objectives for farming tilapia and catfish in ponds over the next five years. This will be complemented by a statement of surface areas exploited, land, labour and capital requirements, research, training, and other necessary inputs and, in particular, who is responsible for what and when.

As already noted, the main feature of a project is that it has a precise beginning and a precise end. In between, a project has a development phase, during which new activities are started, an operational phase, which is reached when these activities are functioning at their planned level (or as near to it as possible) and, finally, a completion phase which includes an evaluation of the results of the project. Thus, in the illustration above, a project in the small-scale fish-farming policy might perhaps be in a second development phase in region A (increasing existing production by 25% by expanding the area under production from 400 to 500 ha), or it might be concerned with the introduction of small-scale farming in region B, where it has not existed before, with 500 ha put into production with the same average yield targeted in region A.

Organization for Project Formulation and Management

Project organization grows in complexity with project size and diversity. There are two organizational options independent of these factors. They are:

Option A: The project sponsor has project formulation and implementation capacity; thus it can:

- undertake complete responsibility for project formulation and implementation;
- partially delegate responsibility; or
- create a special task force.

Option B: The project sponsor does not have project formulation and implementation capability, thus it has to delegate responsibility to a specialized company or agency.

Complete management responsibility is feasible when the project sponsor, such as a government ministry or private company, has its own technical studies and engineering offices, and construction and operational management units, all familiar with the type of project proposed. The project can then be integrated into its regular organization and management operations. This model frequently exists in the agriculture sector, where public or semi-public development boards or parastatal authorities may have such capabilities, but it is still exceptional in the aquaculture sector.

Project organization through partial delegation is more frequent in the aquaculture sector. In this situation the project sponsor may be able to formulate, implement, and operate a project, but lacks the specific technical knowledge required. It acquires this knowledge through the use of independent consultants, or consulting firms, and issues contracts for engineering and all other specialized work required. In this case, the consultants are advisers only, and contractors carry out the instructions of their contracts; all key decisions and management responsibilities remain with the project sponsor. Many capital investment projects in the private sector, and most technical assistance projects financed through international agencies, fall within this category.

A project task force, created within the organization of the project sponsor, is characteristic of large development projects within the public sector, or investment in new fields of activity within the private sector. During project formulation, or at the beginning of project implementation, a specific project task force is established from existing staff which is assigned to the project for its duration, together with additional staff recruited from outside as required. The task force is usually concerned with handling "total packages" in national or international contractual tenders for activities such as project engineering, construction works management, and technology transfer, etc. For projects in the public sector, particular administrative rules have to be followed, ensuring that monitoring, management decisions, and operations levels are independent of each other.

Where the project sponsor does not have any capabilities in project formulation, implementation, and management, it has to contract these functions to a specialized company. In this case the project sponsor limits its role to financing and supervision of what can become full "turnkey" contracts, which may be extended to include not only project formulation and implementation but also management supervision. Such arrangements are similar to those used, for example, by some hotel management companies which build and operate hotels they do not own. It is a relatively new concept in the aquaculture sector, but in the private sector such turnkey contracts are increasing and are now included in the strategies of development banks.

Stages of the Project Cycle

Many international assistance institutions distinguish between five stages in the cycle of existence of a project, namely, identification, preparation, appraisal and agreement, implementation, and monitoring and evaluation. These are shown in the schematic representation (Figure 1) of the project cycle.

At project identification the project idea is translated into a preliminary description of the project. Terms of reference for the project reconnaissance team are established, analyses of existing situations are performed, a broad evaluation of the future "with" and "without" the project is made, and the extent and limits of the project are proposed. Different approaches to the project are identified, and a judgement made regarding which option should be taken forward to project preparation.

At project preparation the project is designed. Objectives, prerequisites, inputs, outputs, organization, participants, clearances are all defined, costs and earnings are calculated, a financial plan is prepared, expected results are analysed, the socio-economic and environmental impacts are estimated, and the provisional and final project documents are prepared.

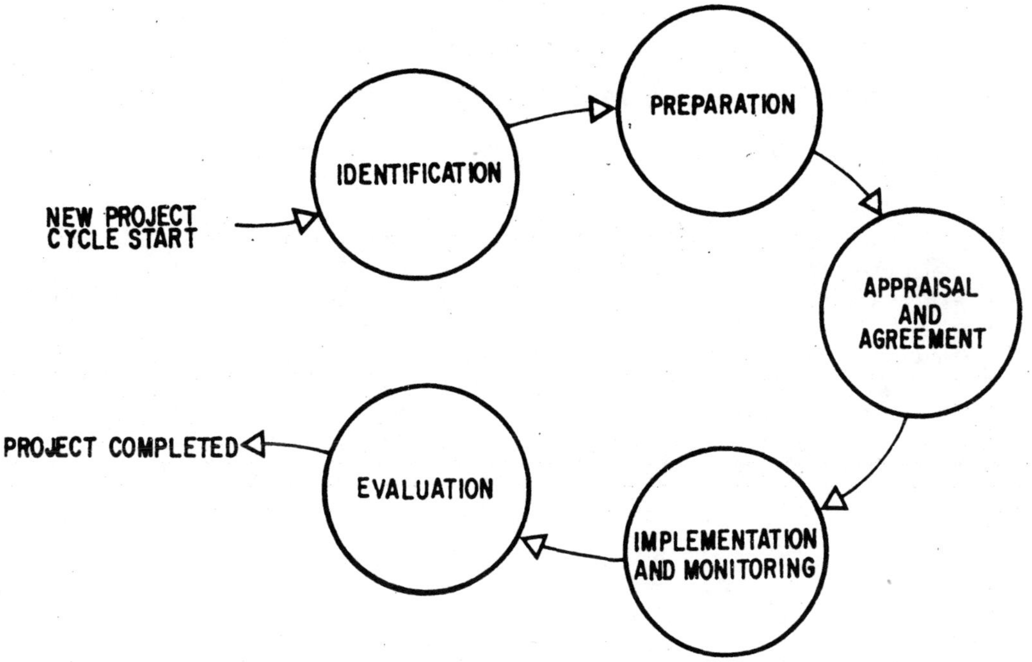

Figure 1 Schematic diagram of the project cycle
Adapted from "Guide for Training in the Formulation of Agricultural and Rural Investment Projects", FAO 1986

At project appraisal and agreement appraisal documents are prepared from the project documents and a succession of appraisal meetings, clearances, and financing negotiations take place. This brings the project to the point of meeting the required start-up agreement conditions, sometimes after revision and adaptation of project schedule, cost, objectives, and financing.

In project implementation the project management and lines of command are established, and various implementation procedures established. In the course of implementation project progress is monitored, revisions and adaptations are made for unexpected events, and finally the project is brought to completion.

At project evaluation, which takes place at a suitable time after the project has been implemented, project objectives, project implementation, and project benefits are appraised. This evaluation may result in the project being extended or in the identification of a new project, and may lead to a revision of the method(s) by which similar projects will be formulated in the future. There are other types of evaluation which may be carried out earlier in the project. Frequently, for example, the FAO Investment Centre will make a "Project Completion Report", or "PCR", of projects which it has formulated. The PCR, which is usually carried out while the project is being implemented, seeks to identify problems which have arisen during implementation and which could have been foreseen at the time the project was designed. Particularly in the case of technical assistance projects, many donor agencies carry out "mid-term reviews" of projects which may result in changes being adopted.

2. AN OVERVIEW OF PROJECT FORMULATION

The Project Idea

All projects start with a project idea. Anyone or any organization which has a reasonably well formed idea about how money and other resources may be invested, may be said to have a project idea. A project idea might be expressed in national sector development plans, in reports of donor agencies' programming missions, in the mind of an individual entrepreneur or farmer, or as part of a private corporation's development strategy. If an entrepreneur or a corporation develops a project idea, financial resources are usually available to carry out further preliminary investigations. Others have to convince financial institutions of the validity of their project idea to carry it further. However, in many cases, experience in aquaculture project planning is lacking.

A project idea may have been developed from research and analysis of extensive materials, and a close knowledge of existing aquaculture activities in the proposed project area. For example, a plantation company may own land, adjacent to a river, no longer required for cane sugar and, after some research into the subject, decides to examine the feasibility of developing it for aquaculture. Alternatively there may be little background to the project, and the idea itself may not be related to any specific species or geographical location. For example, a company in seafood retailing might seek to secure future supplies of high-value fish by investing in farmed production but does not, at that stage, have any knowledge as to where and how an investment would be most profitable.

Whatever the background may be, if it is decided to carry the idea further and to formulate a project, then a process is initiated which takes the project through the first three stages of the project cycle described in Chapter 1, namely, identification, preparation, and appraisal.

Throughout the formulation process, from identification to appraisal, the project is neither approved nor established; it simply obtains increasing degrees of recognition by all parties who may be concerned or affected by it. During this process, many objections and criticisms may be received, and the project's objectives, budget, and schedule may have to be raised and perhaps even restated before the project is accepted. The project may indeed be cancelled at any one of these stages if it does not satisfy the criteria which have been set. Only the last stage, appraisal, leads to the final decision whether to implement the project, with full agreement to its components, or not to proceed further.

The Six Phases of Project Formulation

The three stages of the project cycle are undertaken in sequence, but the work can be broken down in many different ways. The constituent parts may be called "phases", which are further broken down into "steps". Phases and steps can overlap to some degree. Each step contains one or more "tasks".

A framework for project formulation, adapted from the FAO "Guide for Training in the Formulation of Agricultural and Rural Investment Projects", is shown below.

Stage of Formulation	Phase
Identification	Preparation for project formulation Reconnaissance and preliminary project design
Preparation	Project design Analysis of expected results Project documentation and submission
Appraisal	Negotiating the project

Figure 2, also adapted from the FAO "Guide", illustrates this concept and shows the possible overlaps between phases. The activities in each phase are described in the following section.

The framework differs from that often presented which conforms with the general practice of donor agencies where the project formulation team is not usually involved in appraisal, this function being carried out by a separate group, often led by a staff member of the agency concerned. Outside the system of official external assistance, however, it is usual for the project formulation team to continue to be involved until the implementation agreement is signed. Under this arrangement, both project appraisal and negotiation are included in the process.

The phases, steps, and outputs of the project formulation process are shown in Figure 3, and are described in the text of the following sections. This presentation of project formulation is also adapted from the FAO "Guide".

PHASE I: Preparation for Project Formulation

This first phase concerns all activities necessary to prepare for sound formulation of the project. It has two steps:

Step 1 - project inception, and

Step 2 - preparation of a formulation workplan.

The output is the programme of work for project formulation.

PHASE II: Reconnaissance and Preliminary Project Design

The second phase contains all activities necessary to define the objectives of the project, identify and consider options for meeting these objectives, making a preliminary assessment of the content of the project and its likely effects. It normally has four steps, namely:

Step 3 - analysis/diagnosis of the situation from an overall perspective;

Step 4 - analysis/diagnosis of the situation from the perspective of the main interest groups involved;

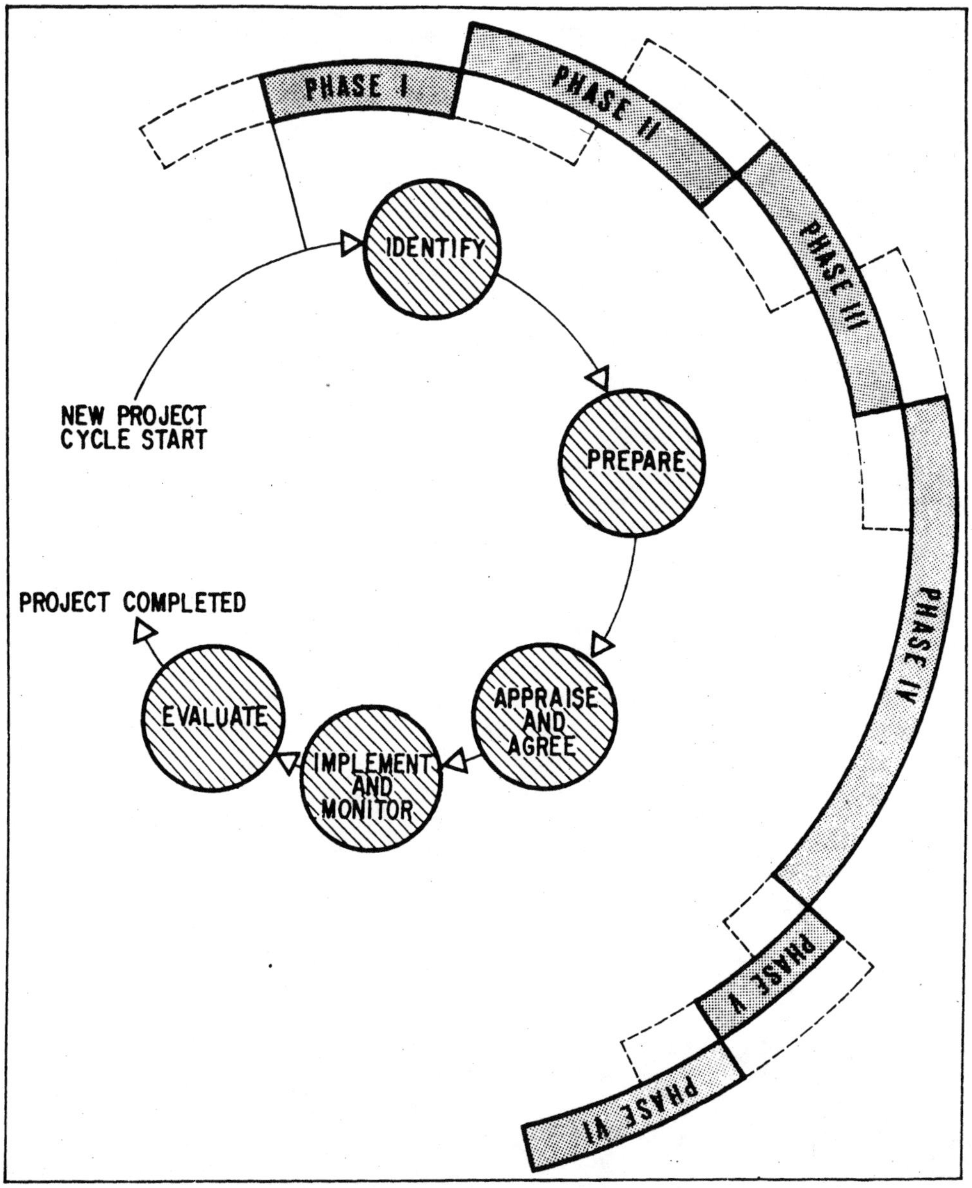

Figure 2 The phases of project formulation and project stages Adapted from "Guide for Training in the Formulation of Agricultural and Rural Investment Projects", FAO 1986

Phase I: Preparatory Organization
Phase II: Reconnaissance and Preliminary Project Design
Phase III: Project Design
Phase IV: Analysis of Expected Project Results
Phase V: Project Documentation and Submission
Phase VI: Project Negotiation

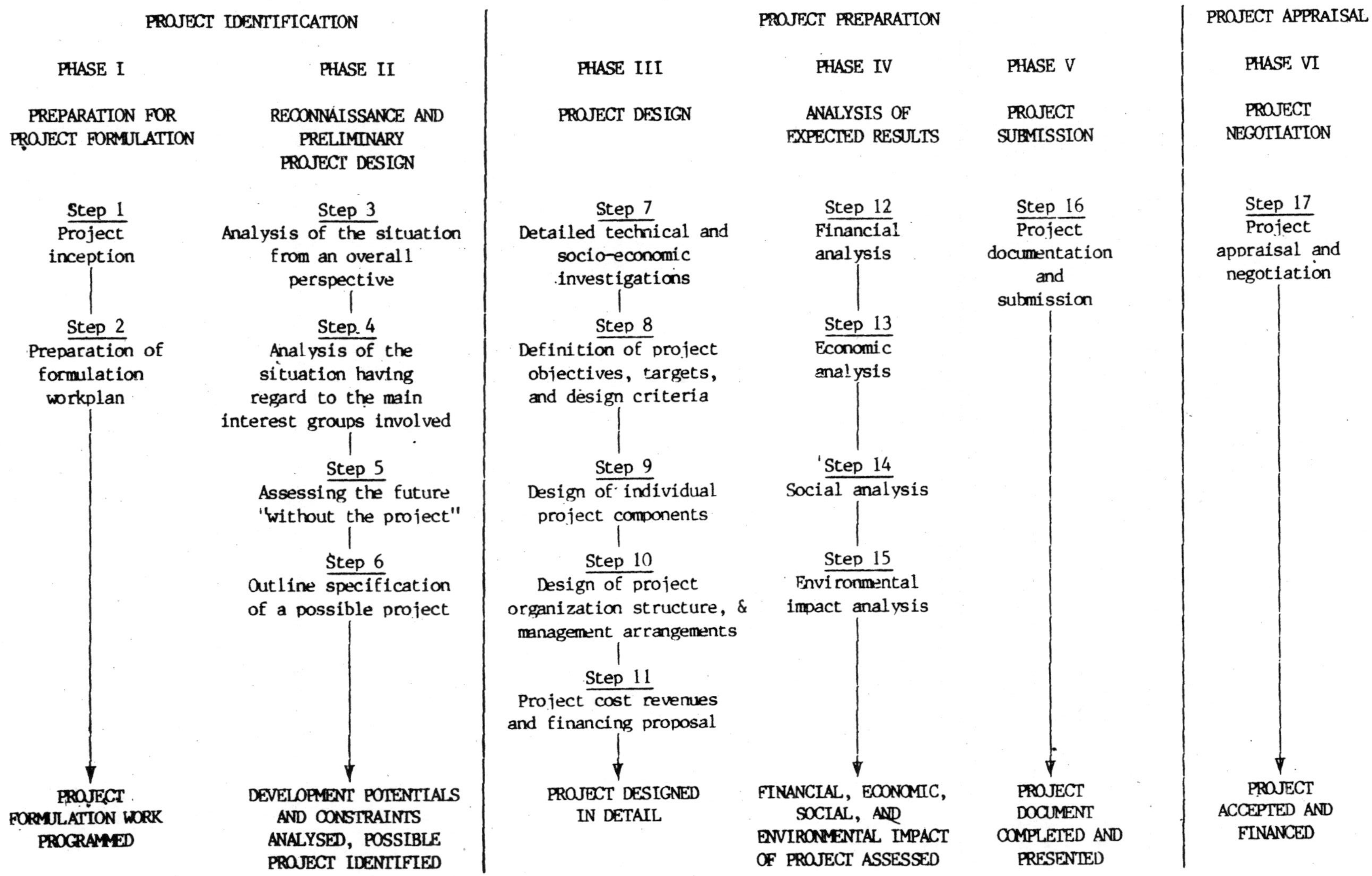

Figure 3 Phases, steps and outputs of the project formulation

Step 5 - assessing the future without the project; and

Step 6 - outline specification of a possible project.

The output of the second phase is the preliminary design of a project, including identification of its main features, such as location, type of participants, main activities, size, timing, organizational structure, and management system. It also includes a first estimate of cost, and a tentative assessment of viability and risk. While this report may be best termed "the identification report", it may also be called a project reconnaissance and preliminary design report. In the private sector it is often called a project prefeasibility study.

At this point the report is normally submitted to the funding organization for approval before further formulation work is undertaken.

PHASE III: Project Design

This phase normally initiates the formal project preparation stage. It typically follows five steps, namely:

Step 7 - detailed technical and socio-economic investigations,

Step 8 - more precise definition of project objectives, targets, and design criteria,

Step 9 - design of individual project components,

Step 10 - design of project organization, structure, and management arrangements, and

Step 11 - project cost and revenues estimation, and first financing proposal.

The output of the phase is a full description and costing of the project, together with a proposed financing plan.

In external assistance projects this phase is not usually the subject of a separate report. In the private sector it may be, when it is usually called a "feasibility study".

A number of adjustments and possibly revisions in project design will often be requested. These lead to additional work, often carried out in parallel with the next phase.

PHASE IV: Analysis of Expected Results

The fourth phase concerns all activities necessary to assess project results in terms of output, effects, and impact on the sector, and on any other sectors they may affect in some way. Its starting point is usually the project design report, which provides the basis for the analysis.

The work typically contains four steps, namely:

Step 12 - financial analysis,
Step 13 - economic analysis,
Step 14 - social analysis, and
Step 15 - environmental impact analysis.

The output of the phase is the determination of effects and impacts of the project.

PHASE V: Project Documentation and Submission

The fifth phase concerns all activities necessary to prepare a final project document, complete with design and relevant analysis.

The work of the fifth phase may be described in one step. It is:

Step 16 - project documentation and submission.

The output of the phase is the project document.

In external assistance projects the work of the formulation team is usually complete at the end of this phase. In the private sector, however, participation of team members may be of value in the project negotiation phase, particularly when several financing agencies are concerned, and/or if negotiations necessitate adjustments to the project.

PHASE VI: Negotiating the Project

The sixth phase concerns all activities necessary to have the project document accepted, and the project financed for implementation. It starts when the the source of financing accepts the project formulation document.

The work includes only one principal step, namely:

Step 17 - project appraisal and negotiation.

The output is a project fully ready for implementation, under proper administration, and with the necessary financial commitments.

3. AN OVERVIEW OF PROJECT IMPLEMENTATION

This chapter describes two aspects of project implementation, namely, the sequence of main implementation phases, and problem areas at different stages of implementation.

Implementation is described by further succession of phases, each phase containing a number of steps. However, because projects differ considerably, the process described below is of a general nature.

In the text, to avoid any confusion when cross-referencing to the work of different phases and steps, phases continue to be numbered consecutively with those in project formulation.

Project Phasing

There are typically up to six phases in project implementation. These are:

Phase VII Recruiting the Human Resource.

Phase VIII Studies and Engineering.

Phase IX Construction and Procurement.

Phase X Start-up of Field Operations.

Phase XI Standardization of Field Operations and Achievement of Project Goals.

Phase XII Termination of the Project Component.

Projects need not necessarily include all these phases, or all in their entirety. For example, in a mussel production and marketing project consisting of four components: a market study for increased production, an extension service for new farmers, a credit programme, and project management, only five of the phases would be necessary. In this situation, Phase IX (Construction and procurement) would not be required, while Phase VIII (Studies and engineering) would consist of the market study, and no engineering would be required. Also the phases would vary in the length of time they occupy in project implementation. For example, project management will begin when the project starts and will finish when it ends, while the market study might be fully implemented early in the project schedule.

The characteristics of each phase can be described briefly as follows:

PHASE VII: Recruiting the Human Resource

This phase begins as soon as the project agreement is signed. Key steps in the phase are:

- recruiting the project manager and project assistants, and recruiting managers and assistants for components;
- establishing a project management office and (if needed) component offices, and providing them with means of operation; and
- recruiting project consultants and (if needed) consultants for specific components.

This phase ends when the last recruitment has been made.

PHASE VIII: Studies and Engineering

The next phase, concerned with project studies and engineering, begins as soon as the project management office is operational. Depending on particular component schedules and characteristics, some studies and engineering work may be delayed (to start-up of the component concerned), or reduced to studies only (if no engineering is needed, as in a marketing component), or even omitted (if no study is needed, as in a training component the complete characteristics of which have been fully defined in project formulation).

Key steps in the second phase are:

- site studies and topographic surveys;
- monitoring environmental parameters;
- engineering of facilities and preparation of tender documents; and
- selection of equipment and supplies, and preparation of procurement lists.

The phase ends when the last construction or supply tender/procurement document is fully completed and approved.

PHASE IX: Construction and Procurement

The phase concerning construction and procurement begins as soon as the first tender or procurement dossier has been approved.

Key steps in this phase are:

- selection of tenderer/supplier and issuing of tenders/supply requests,

- analysis of offers and award of contracts,

- supervision of contractors, works meetings, payment of contractors,

- works and supplies control and agreement, customs clearances, and

- post-delivery control and guarantees.

This phase ends when the last construction or delivery is completed or received, accepted, and terms of guarantees have been met. In construction work this usually means a full year after construction is completed.

PHASE X: Start-up of Field Operations

The next phase deals with field operations. It begins as soon as the first field team has been hired, and provided with necessary working tools, such as vehicles, equipment, facilities, etc.

Key steps in this phase are:

- field testing of all working tools;
- meeting expected minimum biotechnical performance standards;
- meeting expected minimum financial performance standards; and
- meeting expected minimum training, extension, and other standards.

This phase ends when the field operation has reached all the expected standards.

PHASE XI: Standardization of Field Operations and Achievement of Project Goals

This phase begins when project start-up has been completed and management can turn its attention to the main project priorities, such as selling products from the completed ponds, and/or improving the performance of project components.

The phase ends with project termination.

PHASE XII: Termination of the Project Component

This is the last phase. It overlaps the preceding one and begins when all the goals have been reached, and/or funds are exhausted.

Key steps in this phase are:

- preparations for project termination,
- project follow-up, and
- staff reallocation.

The phase ends when the last administrative structure of the project has been closed down, and all staff dispersed.

Potential Problem Areas

The way in which problems occur vary from project to project but they fall into certain well defined problem areas. These may be summarized as follows:

Institutional problems can arise when authority, responsibility, or working relationships are not fully clarified, with formal and properly applied lines of communication and proper decision-making authority among the project team and the organizational units. For example, difficulties in obtaining land for production ponds for a large number of farmers may be the result of dealing with a wrong organization, poor project management and staffing, procurement difficulties, and poor monitoring and evaluation.

Conceptual problems arise from inadequacies in project formulation, resulting from insufficient background studies, or errors in them, or inadequacies in the planning work to be performed. For example, the project may be too inflexible, the objectives may be wrong or non-sustainable, project components may be too many or too large, or the schedule may be unrealistic.

Technical problems arise from unexpected factors discovered during project implementation, or errors in project implementation work. For example, there might be a consistent shortfall in production from fish ponds, due to seasonal water quality changes or poor quality of feed not foreseen in project formulation, or the standard of engineering design and/or construction may have been poor.

Financial problems arise when procedures and schedules for funds, manpower, supplies, and equipment, etc., necessary for carrying out project activities, have not been adequately organized, or delays in implementation result in additional costs, or project costs have been under-estimated. For example, there may be recurrent budget shortages due to poor delivery scheduling of heavy earth-moving equipment for digging new canals or dredging, or, as a result, underestimated investment or operational costs. Also, financial problems may occur because of a change in price for the product as a result of competition or other market factors.

Social problems result from inadequacies in the analysis of social aspects of the project in formulation work, or from changes in social balances/organization during project implementation. Problems may be manifested by, for example, a slow adoption of project techniques by the target group which may find fish farming unattractive or difficult, or there may be an inequitable distribution of benefits as some farms are naturally more productive than others.

Political problems result from changes in national policy or government, or sudden unexpected political events. Insufficient government commitment may result. For example, there may be a change in government to one which is opposed to financial incentives for farmers and all subsidies and grants are cancelled, or a project may be disrupted as a result of internal strife.

Environmental problems often arise and affect the project, and/or the project itself may cause environmental damage which is unacceptable. In the former case these may arise either from project-related or external natural factors unforeseen during project formulation, or from other projects which did not exist or were not planned at the time of project formulation. For example, there may be sudden water pollution from new industrial projects nearby, or competition for the same water resources by urban development; there may be a general degradation of resources, or a natural disaster, such as earthquake or hurricane/typhoon. In the second case, examples may be the heavy discharge of nutrients from the system causing algal blooms dangerous to a nearby oyster fishery, or scouring may be destroying more mangroves than initially foreseen and accepted.

Other problems may occur which may be related to the management and operation of the project itself. These may include the human factor, where, for example, the personality of one member of management team causes minor difficulties to be exacerbated; or there may be a "force majeure".

While it is the concern of the project formulation team to design the project to minimize all these potential problems or risks, so equally the project management must anticipate them as far as possible and minimize their impact when they occur.

It will be apparent from this list of potential problem areas in project implementation that many problems which arise can be foreseen at the time of project design. Satisfactory project implementation, therefore, is dependent on sound project formulation.

Keeping the project on schedule is the responsibility of the Project Manager or, in a large project, the Project Management Office (PMO). Effective monitoring of the project (Step 10) will assist project management in this task. In a small project the project manager will carry out his own monitoring, but a large project may justify formation of a separate group, such as a Project Monitoring Unit (PMU)

Maintenance of the project implementation schedule is particularly important as, more frequently than not, the direct consequences if it is not kept are cost over-runs. It is not an easy task to keep within predetermined time limits. In an analysis of problems associated with some 70 agricultural investment/development projects it prepared in the 1970s, the FAO Investment Centre reported that 80% had over-runs of more than six months, and 40% had over-runs of more than 20 months[1/]. No similar studies exist for aquaculture projects but it is highly probable that the results would be similar.

1/ The design of agriculture investment projects: lessons from experience, 1989, FAO, Rome

There are a number of reasons for the many over-runs in project implementation. Among the most frequent is the considerable pressure on project formulation teams, especially in commercial projects, to complete the project in advance of schedule. Often the teams do not have the capacity to resist 'directives' from a corporate president or a Minister. For example, in a well-intentioned attempt to complete an aquaculture project in time to accommodate a breeding cycle of the finfish or shellfish being cultured (or to lose the whole year), there is a temptation to a project formulation team at the project preparation stage to accelerate the engineering design study or shorten the construction phase, in response to the promise of a Minister or Governor that priority will be given by the Public Works Department for equipment and materials for the project. However, it is less simple or advisable for the implementation team to take such shortcuts in practice. Design and construction proceed at a reasonably uniform and standard rate (unless the project is highly complex) relative to the cost. Without any additional human and financial resources to meet an accelerated schedule, any attempt to do so invariably backfires, causing a delay and a cost over-run, and often disharmony among partners.

Delays can be frequently traced back to a combination of conceptual, management, and technical problems. Such difficulties again often have their basis in weaknesses in the design of projects, particularly failure by the formulation team to ensure that tasks to be achieved in project implementation are compatible with skills and experience of project management, and time is available to carry them out.

Even when implementation schedules allowed for in project formulation are adequate, delays and over-runs may occur due to other reasons. A very frequent problem is the inability of borrowers to comply with conditions attached to the loans.

It can help managers to avoid or counter the problems identified above if they are familiar with the appropriate management tools, such as <u>task analysis</u> and, for larger projects, <u>network analysis</u>.

4. ILLUSTRATIONS OF DIFFERENCES IN THE FORMULATION OF AQUACULTURE PROJECTS

The three previous sections of Part I provided an introduction to projects, together with overviews of their processes and constituent parts. Reference has been made to the fact that, regardless of the type of project and whether it is within the public or the private sector, procedures are similar, even though, depending on the nature of the project, certain steps and even phases may be omitted or the project team may be able to deal with them quickly.

In practice, projects differ significantly and formulation has to take full account of the factors which are special to each situation. Illustrations are presented below of how formulation and implementation methodologies were applied to three projects. All are in the public sector funded through external assistance, but are very different in nature.

<u>A Shrimp Farming Pilot Project In Senegal</u>

This was a single-component project, developed and financed through bilateral cooperation, to construct and operate a demonstration farm to test the feasibility of shrimp farming in a specific environment.

The project was identified in 1980 by a study of aquaculture potentials of the Senegalese coast. The study showed Casamance to be a suitable region for shrimp farming in terms of the availability of sites and ambient temperatures, but concern was raised over hypersalinities of the Casamance River estuary during the dry season as a result of drought in the Sahel. Hypersalinities, together with seasonal variations of temperature, created a pattern of salinity and temperature variations not experienced in other shrimp farming areas. Moreover, there was no existing experience of cultivating the local shrimp species. There was a need, therefore, first to determine if the hypersalinity and high temperatures would be a constraint to shrimp farming and, second, to compare production performances of indigenous species with imported species for which cultivation techniques were known.

Within a few months preparations for project formulation were arranged between the Government of Senegal, a French overseas development agency, and a French consulting firm. Reconnaissance and preliminary project design were considered as having been, in practice, already carried out in the course of the original study. The project formulation team went, therefore, straight to project design, including a full techno-economic study in the format required by the agencies concerned. The terms of reference were to formulate a test to demonstrate, in conditions typical of the area, the technical and economic feasibility of shrimp farming in the large available area within the Casamance delta. The test was to be located at a site representative of average conditions in the region, from which results could be extrapolated to the area as a whole, rather than at a site particularly favourable to shrimp farming in terms of salinity and temperature ranges.

In order to minimize investment costs, test facilities were limited to 15 000 m^2 of earthen ponds. It was assessed that no hatchery would be required, as shrimp post-larvae would be imported from abroad, and juveniles of native species would be collected from the wild.

The process of project formulation, appraisal, and negotiation was completed in less than a year, and the implementation agreement between the donor agency and Senegalese Ministry concerned was finalized in 1982. Implementation was begun immediately. The project, as formulated, was a single-component project planned for implementation through a turnkey contract, with international procurement restricted to French suppliers.

Within a short time the phase of recruiting the human resource was completed, a project manager having been assigned by the Ministry, and a two-year management contract having been signed with the French firm which had performed project formulation work. The phases of studies and engineering, and construction and procurement were executed as planned. The phase of start-up of field operation, however, faced considerable difficulties in the procurement (from Latin America and Asia) of the post-larvae of non-indigenous shrimps which were needed in the required quantities and qualities, and at the proper times. This resulted in delays in performing the growing tests as planned. Consequently, when the two-year test period came to an end, the project, instead of having clearly demonstrated results, had only preliminary data for six shrimp species, of which four were giving promising results.

It was decided that there should be an extension of the test project, with use of the original ponds to establish stocks of captive spawners, construction of a small hatchery, and of four new ponds of 1 ha each. This project extension, with allocation of additional funds, had new phases of construction and procurement and start up of field operations.

Finally, in 1986-87, the project was able to supply most of its post-larvae needs. An annual production of between 2-3 t/ha could be achieved with two production cycles a year, using different species of shrimp in each season. However, target levels of efficiency in spawning and hatchery operations were not reached, and repeat demonstrations could not be carried out in several ponds.

Full-time technical assistance was withdrawn at this point, the project being terminated without having reached the implementation phases of standard field operation and effective achievement of project goals.

A second project was then started, based on an external evaluation of the first. The purpose of this project, through the continued operation of facilities built in the first project, was to study ways of encourging investment in shrimp farming in the area, and of developing the sector generally.

A Shrimp Culture Project in Bangladesh

This multi-component project, funded by the World Bank, had the objective of improving shrimp yields through improved water management, introduction of new production practices, and a loan programme for farmers.

Discussions between the Government of Bangladesh and the Bank regarding cooperation in coastal aquaculture started in 1979 and resulted in identification of a shrimp culture project in 1981. Project formulation and negotiation took almost five years. The project formulation report was completed in mid-1983 and discussions on the basis of the report extended for another year. An appraisal mission was undertaken in late 1984. Credit negotiations took place in late 1985, and the final staff appraisal report was delivered at the end of 1985. The implementation agreement was signed in mid-1986.

Project implementation began that year. Recruiting the human resource extended until mid-1987, with establishment of a Project Management Office, recruitment of national staff, and recruitment of foreign consultants through an international tender procedure.

At the time of implementation, the project was one of the largest in the aquaculture sub-sector, in terms of budget (over US$ 36 million) and number of target beneficiaries. It concerned small-scale farmers in about 7 000 ha of low-lying coastal lands already used for shrimp culture in the Districts of Cox's Bazar and Khulna. In this area the project provided an extension service, credit, and public investment. In another area of 2 500 ha, the project provided extension and credit only. The project had four components, namely:

(a) Infrastructure: Design and construction of embankments, water control structures, and shrimp hatcheries (including two demonstration hatcheries in the public sector), to provide more efficient management in the main 7 000 ha project area, and to supply high quality shrimp seed.

(b) Credit: Supply of medium-term credit to shrimp farmers, traders, and hatchery operators, for financing on-farm investments, marketing equipment, and private hatchery investments.

(c) Institutional support: Financing of buildings, equipment, incremental staff salaries, and incremental operating costs of the Department of Fisheries and the Bangladesh Water Development Board, to strengthen their ability to implement and monitor the project.

(d) Technical assistance: Consulting services to assist the Project Implementation Unit (PIU) in project implementation, training, and coordination of project activities, and to assist in training shrimp farmers, and in providing fellowships for education/training specialists and PIU staff.

The project cost was estimated to be US$ 36.7 millions, of which 60% was from the World Bank, 12% from the United Nations Development Programme, 12% from the Government of Bangladesh, 7% jointly from the Bangladesh Krishi Bank and the Sonali Bank, 5% from private Bangladeshi investors, and 4% from the Bangladesh Bank.

The project was one of several fisheries, aquaculture and irrigation projects in the country, many of which had already been completed (such as the Freshwater Aquaculture Development Project funded by the World Bank, and the First Aquaculture Development Project funded by the Asian Development Bank), or which were being implementated (such as bilateral technical cooperation projects with France and the Federal Republic of Germany), or which were being negotiated and would start soon after (such as the Second Aquaculture Development Project funded by the Asian Development Bank). These projects were essential to the development of this key sector in the country for food production for the domestic market and generation of export revenues. All were particularly targeted toward the rural poor.

By 1990, some delays had been encountered and adjustments made to the contracted work plan but, generally, it was in line with the original project concept.

The ASEAN Aquaculture Development and Coordinating Project

This was an interregional aid project, funded by the European Economic Community and ASEAN countries. It was concerned with the provision and chanelling of technical assistance for scientific cooperation and technical training in aquaculture, and involved the twinning of institutes in Europe with institutes in the ASEAN region.

The project concept dated from the beginning of the 1980s, with a proposal from the ASEAN Committee on Food, Agriculture and Forestry to the EEC. Project formulation and negotiation was undertaken in 1985-86, with a project formulation contract between EEC and a Dutch University-related consulting group. This led to project approval in November 1986.

The formulated project had objectives of (a) strengthening applied research, development, and training services of selected aquaculture institutes within ASEAN, thus helping to upgrade technical capacities within the region; (b) improving socio-economic conditions of the rural poor through aquaculture development, specifically through increasing supplies of animal protein, increasing income, and providing additional or alternative employment opportunities; (c) optimizing available marine, coastal, and inland resources in ASEAN for aquaculture; (d) initiating long-term ASEAN regional collaboration and technological communication, and (e) fostering long-term collaborative linkages between EEC and ASEAN institutes engaged in aquaculture development, training, and research.

The project, coordinated from Thailand, had programme components in selected institutes in all five ASEAN countries, each component being the theme of a joint work programme by twinned EEC and ASEAN institutes.

In addition to the Project Management component, the five sub-sectoral components were:

Indonesia - to improve methodology of coastal zone assessment for aquaculture, including site surveys, and technological, ecological and socio-economic studies.

Malaysia - to improve methodology of assessing lakes and reservoirs for fish production, including survey technology, and the development of cost effective management systems.

Philippines - to improve capabilities of ASEAN training specialists in aquaculture, and develop applied aquaculture technology and training.

Singapore - to develop feeds for marine fish and transfer marine culture and post-harvest technology.

Thailand - to undertake applied research in fish and crustacean genetics to hasten domestication and increase production through development of faster growing, disease resistant strains.

On the ASEAN side participating institutes were nominated by their governments. On the EEC side, however, the procedure preferred was that of international tenders issued to a long list of interested institutes, evaluation of their capabilities by a committee of EEC experts, and establishment of a short list of institutes to carry out each component.

The project inputs included:

- Staff provision and exchange: 200 man-months of technical assistance and other professional services to ASEAN institutes, 250 man-months of staff exchange between twinned institutes, and 88 man-months of staff exchange between ASEAN institutes themselves.

- Interregional seminars: Two EEC/ASEAN seminars per component (one in the country of each twinned institute), plus two general interregional programme workshops (one in ASEAN, one in Europe).

- Training: short-term courses for each component totalling 600 man-months of training, plus two long-term overseas scholarships per component for ASEAN staff.

- Equipment: miscellaneous communication, scientific, and laboratory equipment, technical equipment (for ponds, cages, hatcheries, harvesting, feed preparation), and a small number of boats and vehicles to the ASEAN institutes.

Implementation of this five-year project began in late 1988 with signature of the general project agreement between the European Development Fund of EEC and the Committee on Food, Agriculture and Forestry of ASEAN. A Project Management Office was established at the Ministry of Agriculture and Cooperatives (Thailand), and appointments of a Project Manager (from Thailand) and an international expert as Project Coordinator. Effective field operations were scheduled for mid-1990, after approval by the Project Steering Committee in November 1989 of the five component workplans of twinned institutes.

Although not one of the largest development projects in terms of budget allocation (about US$ 11 million), the project was large in terms of geographical area (five ASEAN countries), and in its number of participating institutes (ten institutes from nine countries).

A mistaken assumption in project formulation caused considerable delay in the procedure of selecting international participants by the EEC. When the project was formulated (1985-86), it was assumed that participating institutes would not charge fees for their professional services but only request funds to replace staff on long-term assignment in ASEAN. Adequate arrangements between the different parties which remained interested were not reached until mid-1989. At that time agreements were finalised with, on the EEC side, a British institute (for Malaysia), a Dutch institute (for Thailand), two French institutes (for Indonesia and Singapore), and a French/Italian consortium of institutes (for the Philippines), and, on the side of ASEAN, a number of institutes in the region.

Besides economic, social, and technical benefits to the sector in the region, many of which were anticipated to be long-term and indirect because of the nature of the project, the main benefit is expected to arise from the creation of a framework for future interregional collaboration with the five pairs of twinned institutes setting the scene for future EEC/ASEAN collaboration, in close coordination with other agencies and with programmes undertaken in the region.

PART II. PROJECT IDENTIFICATION, PREPARATION, AND APPRAISAL

1. AQUACULTURE PROJECTS COMPARED WITH THOSE FOR AGRICULTURE

Part I contained an introduction to the project cycle, and illustrations of the processes of project formulation and implementation. Part II now describes in greater detail the stages within project formulation, that is identification, preparation and appraisal

It is important to point out at the outset that production-oriented aquaculture development projects differ substantially from those typical of production in agriculture.

Agriculture production projects are often concerned with large-scale development over wide areas of land for which there may be options for different crops, such as sorghum, corn, maize, tobacco, vegetables, etc. Even production projects which perhaps only deal with one type of farming system, such as animal or poultry husbandry, are often not restricted to any one place.

Aquaculture production projects in general are more constrained physically by their need for existing waterbodies, or water resources adjacent to suitable land. Consequently these basic requirements immediately limit options for:

- the species which can be farmed (into those which tolerate freshwater, brackishwater, or marine environments);

- the systems which can used (as extensive, semi-intensive, and intensive systems all depend on the availability of resources and inputs); and

- the practices which can be used (as all farm units, such as ponds, raceways, floating cages, rafts, etc., each have characteristics which make them particularly applicable in certain conditions).

These physical constraints influence the types of projects which can be formulated. Notably, projects with activities which extend over large areas are, for the most part, not characteristic of aquaculture development; projects are typically relatively small, and highly specific in terms of their objectives.

While the emphasis in this document is on public sector funded projects, it is recognized that in aquaculture, as in agriculture, most investment is made by the private sector, with little or no direct intervention by government. Such investment decisions can be taken within the broad framework of project formulation described here, but the process will differ in the level of attention given to each stage.

As a consequence of the constraints and the limitations noted above, aquaculture projects, by and large, fall into four categories or models, i.e.:

MODEL A: Private sector projects where investment is by a commercial interest. For example, an entrepreneur wants to build a trout farm on his own land; or a shrimp farming company wants to build a new hatchery; or an international corporation wants to convert its low-grade sugarcane plantations into catfish production.

MODEL B: Public sector projects where the investment is in a publicly owned entity. For example, the Ministry of Fisheries wants to build a state hatchery to support its programme to enhance inland fisheries; or the national extension service requires a new farm to demonstrate freshwater shrimp production; or the Ministry of Technology wishes to build a national research and development centre; or a parastatal organization wants to build a new fish market.

MODEL C: Public sector projects where investment is by private farmers supported by government services. For example, the Ministry of Agriculture wants to increase inland fisheries production; or the Department of Aquaculture wishes to increase national production of molluscs; or the Ministry of Planning and Development seeks to increase national foreign exchange earnings through marine shrimp production.

MODEL D: <u>Public sector projects concerned only with institution building</u>. For example, the Ministry of Agriculture needs to be strengthened to improve its organization and management of the aquaculture sub-sector; or a parastatal body needs training in market research and product promotion; or certain universities want to upgrade their graduate and post-graduate education programmes in aquaculture.

2. PROJECT IDENTIFICATION

Project identification is the first stage of formulation. It has two phases (I and II). The first concerns the preparation required for project formulation, and the second concerns reconnaissance and preliminary project design.

PHASE I: Preparation for Project Formulation

The <u>purpose</u> of Phase I is to prepare for well-organized project formulation, or, more precisely, to complete all necessary arrangements (administrative, logistical, and financial) to ensure that project formulation makes efficient use of the time and resources it requires.

Four main <u>outputs</u> of this phase are expected:

(i) The project concept or idea is outlined, with basic objectives, possible activities, and boundaries. The original concept is formalized and described in a written statement on the basis of available information. The preliminary project framework is constructed (public sector projects).

(ii) Terms of reference (TOR) for the exercise are prepared. They determine the broad scope of work to be carried out. They are used, subsequently, as a yardstick to measure effectiveness of the formulation team.

(iii) The work plan for carrying out formulation activities is prepared as far as possible. This requires participation of the team leader and involvement of all team members to ensure that each professional requirement for the work is taken into account.

(iv) Specific individuals required in the formulation team, representing the different disciplines required, are identified and recruited. Funds are released to provide operational expenses, and office space, transportation arrangements, resources for communication, and any other logistical support, are all reserved.

The work in the Phase I typically has two steps, namely project inception (called Step 1), and preparation of formulation work plan (Step 2). Activities of both these steps are illustrated in Figure 4, and described in detail in the following text.

For projects financed wholly or partially through official external assistance, a <u>logical framework</u> is often required at this point. This is a planning and management tool which is complementary to the conventional project formulation procedures which are the subject of this document. Table 1 shows a logical framework, written at the project idea stage, of a project for a municipally owned fish farm in China.

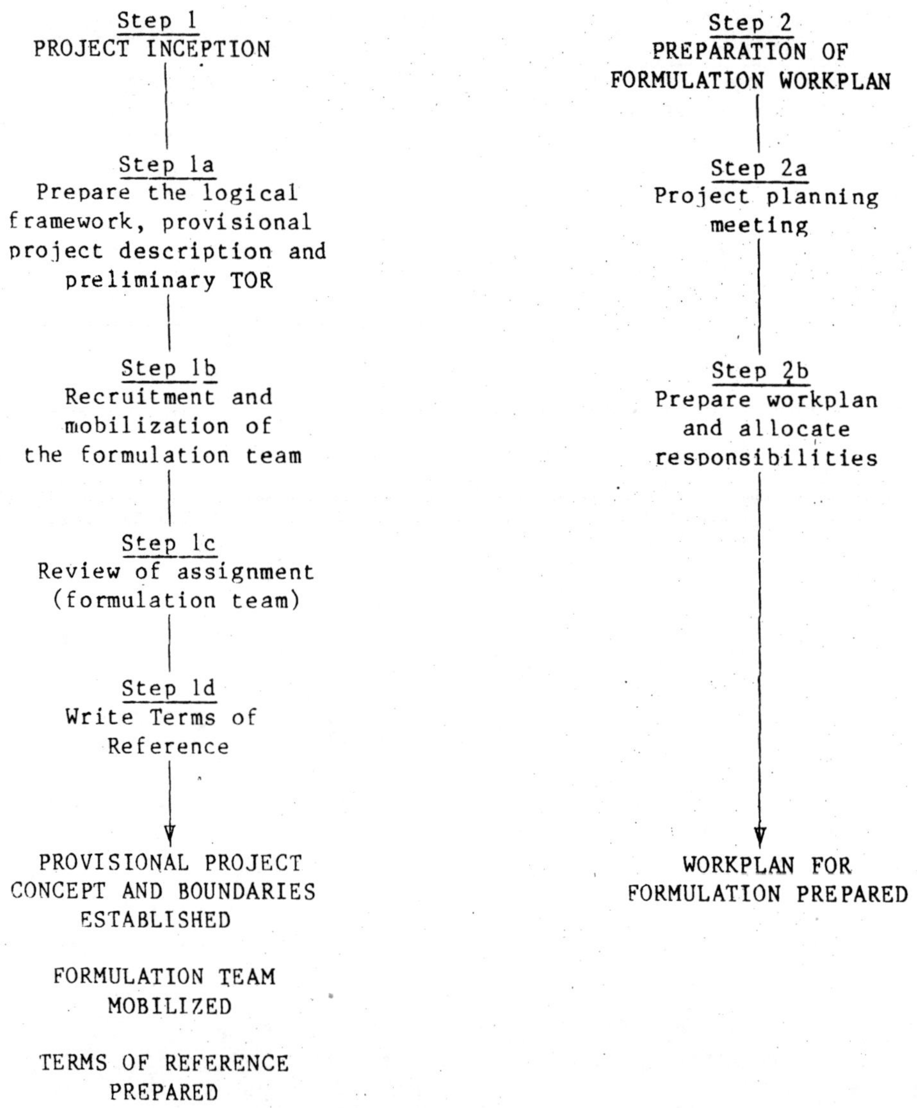

Figure 4 Phase I. Preparation for project formulation

For projects financed wholly or partially through official external assistance, a logical framework is often required at this point. This is a planning and management tool which is complementary to the conventional project formulation procedures which are the subject of this document. Table 1 shows a logical framework, written at the project idea stage, of a project for a municipally owned fish farm in China.

The logical framework is a guide to helping those concerned to "think through" the project by taking them through a number of logical steps. Thus, at the first possible point in the formulation process, those responsible for the project are required, after defining sector objectives, to develop the first draft of the framework. This requires the project planners to:

Table 1

A logical framework at project inception stage, of a proposed project for a municipality, fish farm in China

Narrative summary	Objectively variable indicators	Means of verification	Important assumptions/risks
Sector Objectives/Targets			
Increase freshwater fish production by x tons, for the domestic market	1) Increase in freshwater fish production 2) Increase in supplies of fresh water into the market	Official statistics Official statistics	1) Resources are available 2) There is a market to absorb the increase
Increase employment by x%	Number of new jobs created	Official statistics	Labour of suitable quality is available
Incerease contribution to GNP by 5%	Value of incomes generated	Official statistics	Revenues exceed costs
Project Purpose, Immediate Objectives			
Investment in construction of 100 ha of pond area, hatchery, pumping facility, canals, buildings, access road construction, laboratory	1) Investment capital disbursed on schedule 2) Farm is constructed and commissioned 3) Fish sales	Financial statement (a) Implementation completion report (b) Report by the verification group Sales records	1) Quality of materials and construction are acceptable 2) Capital costs are within estimate 3) Management, technicians and labour of required expertise available
Financially profitable	Profit	Accounting records	Revenues greater than costs (capital and operating)

(Cont.)

Table 1 (continued)

Narrative summary	Objectively variable indicators	Means of verification	Important assumptions/risks
Outputs			
Sales of fish	Quantity sold	Sales records	1) Demand exists compatible with production
Provide on-the-job training for technicians and managers	Number of people trained and moved to new projects	Personnel records at the farm	2) Supply is consistent with estimates of yield and mortality (a) There are managers & technicians available to provide the required training (b) Suitably qualfied people are available for on-the-job training
Job creation	Number and type of jobs	Personnel records at the farm	(a) Managers, technicians and labour of required skills are available
Financial surplus	Profit	Accounts	(b) Successful operation of farming
Inputs			
Land	Government will assign land	Land use certificate	Not suitable for other use
Water	Government will issue water lease/use permits	Water use certificate	Quantity and quality adequate
Access roads Power and services Pond and canal construction Pumps/Buildings/Vehicles	Completion	Acceptance report	Release of funds and work completion are scheduled
Raw material for fish feed	Delivery checks	Invoices/receipts	Materials are available
Trained personnel	Present	Personnel records	Available
Working capital	Bank transfer	Accounts	Available

- set indicators or criteria related to project and sector objectives;
- set means of verification of the project objectives;
- state the major assumptions and risks affecting the project.

At this point in the formulation process, the information available is likely to be at a general level, as shown in Table 1. Nevertheless, those responsible for developing project ideas, or securing funding to take them through formulation, will find the framework useful. As more information becomes available in project identification and design, more detail can be written into the logical framework.

The framework approach is most useful in projects which have a specific and quantifiable output, for example production projects. It is not so convenient to use for wide ranging projects, or where it is difficult to verify objectives, for example in research, training and institutional strengthening projects where the means of verification are primarily qualitative.

With some donor agencies the decision to proceed to project formulation is made on the basis of the logical framework.

Step 1 - Project Inception

Although procedures vary between countries, agencies, and also between the public and private sectors, it is good practice for the project sponsor (a donor agency or investor in the private sector) to draft preliminary TOR. These will be finalized subsequently, preferably by the formulation team in conjunction with the sponsor.

Step 1a - Preparation of the logical framework, provisional project description and preliminary Terms of Reference (TOR)

The starting point of the project formulation process is the project idea. The idea may have originated from any of a number of sources, including a previous project, sector studies, and requests from interested parties. It may be brief and broad in scope, or quite precise, and it may already be linked to a particular governmental department, or a funding source.

When the project sponsor has decided that a project should, if possible, be formulated, it is desirable that a brief project description should be prepared, and preliminary TOR written. It is important, when doing so, to maintain the right balance between, on the one hand, the need to keep the project concept flexible (so that any potentially significant area, sector, institutions, or target group are not excluded *a priori*) and, on the other, the need to keep investigations within constraints of time and resource limitations. For this reason donor agencies usually appoint an experienced staff member to undertake this task.

It is useful when drawing up TOR to set out preliminary options concerning geographical boundaries, institutions which will be involved, and target groups. Based on these options, the preliminary TOR should take into account the following:

- the nature of perceived development problems and priorities attached to their solution;

- broad project objectives and how they fit the overall development strategy (of the region, country, institution, or company);

- decisions already made on main project components, and information on components for consideration by the formulation team; and

- decisions already made regarding specific areas, sub-sectors, institutions, and target groups to be involved, or information for the team to consider.

Step 1b - Recruitment and mobilization of the formulation team

The composition of the team, in terms of professional disciplines and technical or vocational experiences respective for the work required, mainly depends on the nature of the project as described in the provisional project description and preliminary TOR. It is advisable to keep the team fairly small (probably a maximum of five persons, including team leader). The individuals within the team, therefore, must cover collectively a broad range of professional disciplines and expertise.

In aquaculture, because it is a new field, it is not always easy to bring together individuals who have the necessary broad experience and can direct their work immediately to the key issues of the proposed project, and it may be difficult to find the specialists who are also required. For example, for a project which has a component of commercial production, it is important at the outset to include one or more individuals who have worked in the commercial side of the sector for a living (such as a farmer or processor), rather than those whose production experience is only information gained as a by-product of technology research and development. Similarly hatchery engineers, who may be experts with certain types of systems for temperate species, are often not familiar with technology for tropical species, but their knowledge can assist in the transfer of technology.

The nomination of team leader is critical. The team leader should be experienced in at least one of the expected project components, well-qualified to command the respect of all team members, and capable of assembling and writing the project report.

Step 1c - Review of assignment (formulation team)

Using the provisional project description and preliminary TOR as a starting point, the assembled formulation team members should make themselves familiar with the assignment prior to commencing work in the field. This would include:

- familiarizing themselves with all project criteria;

- assembling existing data and documents relevant to the project;

- consulting key representatives of the sponsoring organization, government bodies, organizations, and individuals who possess information or knowledge relevant to the project, or who can express points of interest, or policies;

- identifying from this information essential development issues and problems which should be given priority during the field work in Phase II.

The attention given to this Step varies considerably. It is not uncommon for teams to go into the field immediately after mobilizing, with little or no briefing beforehand or, indeed, to mobilize in the field. This may be because project sponsors which are development organizations may have considerable "institutional memory" immediately available to draw upon and, very often, the team leader will have recent experience of the area to which the team is going. Even when this is not the case, the Step does not require time consuming schedules for data or preparation of new material; simply, what is necessary is the assembly of existing knowledge and its use to identify the major issues. Going through this process will make project identification easier and more effective.

It is important, however, that the team knows all criteria which the project sponsor has set out for the proposed project. These criteria fall broadly into two types, those dealing with objectives, and those with constraints.

Objective criteria concern aims, such as minimizing foreign expense (for procurement of inputs and equipment), achieving a minimum level of profitability, maximizing hard currency earnings (producing export crops), optimizing manpower usage, increasing food supplies, etc.

Constraint criteria concern predeterminants, such as ceilings on total project cost, costs per beneficiary, maximum technical assistance cost, the location of the project, the selection of cultured species, etc.

Step 1d - Terms of reference

These are TOR for the formulation team as a whole. The preliminary TOR (Step 1a) are now redrafted, having regard to the analysis of the existing information carried out in Step 1c, so that they are a useful guide to the team.

Writing good TOR is not a brief and simple task. In many cases poor project formulation can be traced to carelessly drafted TOR. They should not be too brief and vague, thus leaving the formulation team with a potentially large and endless task. Alternatively, they should not be too detailed and specific, leaving the formulation team little room to manoeuvre or to contribute imagination and creativity to the task.

Good TOR for a project formulation team should:

- provide a concise background to the assignment (the origin or rationale behind the project idea),;

- state clearly what is expected to result from the work of the formulation team;

- specify important or general national or sectoral policies with which the proposals of the team should comply or be consistent;

- indicate major project selection and design constraints which should be observed (such as maximum project cost, or target farm incomes);

- suggest types of activities which may be included in the project, and which institutions could be involved; and

- set a precise date for conclusion of the formulation exercise.

TOR should be restricted to no more than a few pages typed in normal format. They should express a manageable task, the achievement of which is realistic within the permitted time frame.

From this point the team is responsible for organizing its own work.

Differences in public/private sector procedures (Step 1)

While all the tasks of Step 1 are common to all four models of aquaculture projects, there are some differences within each of the tasks between public and private sector procedures. Notably, in Step 1a, the construction of the logical framework is usually omitted in private sector projects, and in Step 1c (review assignment) there could be important differences in criteria for a private sector project (Model A), compared with those of public sector production projects (Models B & C). For example, for the entrepreneur who wishes to use his own land and water source for trout farming, the project criteria will include references to these site constraints. For a government wishing to build a new catfish hatchery to enhance a large reservoir fishery (Model B), the species and production required may be established but not yet the exact location of the hatchery.

Step 2 - Preparation of the Formulation Workplan

With the right TOR the tasks of the leader and his team are manageable, and the schedule and objectives are realistic. During this Step the team plans effective implemention of work in the time allocated. Project formulation is always undertaken within limited time and resources. The work, therefore, has to be carefully oriented and scheduled so that the team completes it on time, within the framework of the project criteria, and within budget.

Step 2a - The project planning meeting

The purpose of the meeting is to prepare team members for their tasks. When possible the meeting should be held in the presence of the project sponsor, whether an individual entrepreneur, a company representative, or a steering committee (in the case of most agencies). It provides an opportunity for ensuring that the individual team members fully understand what is expected of them, and accept and adopt a common approach to their assignments (in terms of method and procedure). The "working documents" for the meeting will be the TOR (Step 1d). At some planning meetings important decisions may be taken regarding, for example, scheduling, or the types of project analysis which it is planned to adopt. For the team leader, the meeting should facilitate the preparation of the workplan.

Minutes of the meeting are recorded by the team leader and circulated to the team and the project sponsor.

The planning meeting is often not as rigorous as it should be. In private sector projects it may be no more than a discussion with the client, or company representatives; in the public sector it may be a round-table gathering with the project sponsor and government departmental staff members.

Step 2b - Preparation of the workplan and allocation of responsibilities

In most cases, aquaculture projects require only a simple workplan which can be illustrated in the form of a bar chart of key tasks. In relatively complex projects, where there are many inputs in the work to be undertaken, the team leader may construct a "logic diagram". This shows the relationship in time of different tasks and the total time required. It may then be helpful for him to use a "task allocation matrix" by which various tasks can be cross-referenced with individual team members who will thereafter have responsibility for them, whether working alone or with a group. It is essential that the schedule, description of each task, output required, responsibility for the output, and date of delivery, are all accepted and adopted by the team members at this point.

Scheduling is now completed, and there is a precise list of responsibilities for each team member. The team leader should now prepare internal TOR for each team member based on the task allocated. This prevents misunderstandings with regard to what is expected of each member of the team, and when.

In larger and more complex project formulations it is usually necessary to use "critical path analysis" (CPA), a management technique to achieve a manageable workplan, usually through Programme Evaluation and Review Techniques (PERT).

Differences in public/private sector procedures (Step 2)

Both tasks of Step 2 are common to all four Models (A, B, C, & D).

<u>Illustrations of Phase I: Preparation for Project Formulation (Steps 1-2)</u>

Project 1 - <u>A Commercial Fish Hatchery Project in a Mediterranean Country</u>

This private sector project was initiated by an investor who was already undertaking research and development in cage farming for sea bass and sea bream in the country. After struggling for a few years from an erratic supply of seed, the lack of good feed, and outbreaks of disease, he considered that it should be more profitable to concentrate his business on the production of seed to be sold throughout the Mediterranean region. He had already purchased an area of land adjacent to his site to protect his farming activities, and he believed this would be suitable for a hatchery and nursery.

In discussing his idea with fish culture experts at a Conference and Trade Show (Step 1a), he estimated that a hatchery and nursery producing about 5 million four-month old juveniles each year would be profitable. After visiting a number of consultancy companies exhibiting at the Trade Show, he asked three companies to submit their qualifications and a proposal, with unit costs for professional services, based on written preliminary TOR and background information which he provided (Step 1b).

One month later he received all three proposals in which each company, from available information, identified the key issues as they saw them

(Step 1c), together with a general description of the approach proposed (Steps 2a, 2b, and 2c). He selected the firm which had the clearest TOR (Step 1d) and a contract was drafted for a three-month preliminary site study, followed by separate price estimates for the complete prefeasibility study and then the feasibility study (Step 1d). The consultants estimated the prefeasibility study could be completed in nine months, if local assistance was provided for a topographical survey, soil data, and water quality analysis, and the feasibility study possibly six months later.

The investor met the consultants (Step 2a) to confirm the time requirements and workplan (Step 2b). After three days of discussions a lump sum contract was agreed for the pre-feasibility study. This included a site study, but the investor agreed to supply separately the survey, and soil and water quality analyses (Step 6b).

The phase of preparation for project formulation took under two months.

Project 2 - A Brackishwater Shrimp Farming Project in a Southeast Asian Country

The external assistance project, as provisionally described in the TOR prepared for the formulation team (Step 1a), envisaged the construction and/or rehabilitation of canals and of existing fish and shrimp ponds to achieve higher yields of shrimp, and the construction of hatcheries, in specified areas. This was to be accompanied by a strengthening of research related to shrimp farming, the provision of extension services to fish-farmers, and the provision of credit to farmers.

Formulation was to be done in two stages: (a) individual schemes would be identified and ranked in order of priority for attention, and preliminary proposals prepared for their implementation; (b) "sub-projects" would be designed to a feasibility level and plans would be prepared for the institutional infrastructure (hatcheries, extension, credit, and research) covering all the sub-projects.

The available funding allowed for a team of five experts to have three weeks in the field and three weeks report writing for Phase II. The team consisted of a team leader/economist, and institutions, civil engineering, and shrimp farming and marketing experts (Step 1b). The team leader and two of the other experts had much recent experience of similar work in the country which substituted for a more formal collection of information about the country (Step 1c).

In this mission the redrafting of the TOR (Step 1d) was omitted.

The team, other than the credit expert who had been delayed, mobilized in the capital where a meeting took place attended by representatives of the Department of Fisheries (DOF) - (Step 2a). At this meeting the objectives of the project were confirmed by DOF, the team's itinerary was discussed and agreed, and the roles of the individual team members. The team leader presented proposals of the lay-out of the interim report and the list of annexes which, at that stage, appeared to be necssary.

The constraints of time available and the itinerary largely determined the workplan and, other than identifying what was expected from each team member, it was not considered necessary to go further into detail (Step 2b).

Project 3 - A Shellfish Farming Enhancement Project in Southeast Asia

In a meeting between EEC and a regional intergovernmental body from Southeast Asia to discuss technical cooperation and development, the general area of improving sanitary control in the production and marketing of farmed shellfish throughout the region was acknowledged to be a priority. It was agreed that a team of EEC consultants would tour the region on the premise that Europe had technical knowledge on shellfish farming and sanitary control which could be transferred to improve these activities in the region. Increased production of shellfish in the region could be sold in national and export markets, and new markets would be opened for European equipment for the sanitary and depuration processes.

Six months later a team of three EEC experts (an economist, who was the leader, a shellfish producer, and a commercial processor) toured the region, discussed the idea with the five respective national Fisheries Departments concerned, and prepared the provisional description of a large, multicomponent, ten-year project (Step 1a), to be discussed further after a study tour in Europe by a delegation from the region.

Details of the project were circulated to fisheries institutes in the EEC and the region, and the study tour was budgeted and organized. Ten months later a second meeting was held in Brussels. Project objectives, to help small-scale fishermen re-orientate toward shellfish culture, were redefined and accepted.

An EEC tender was prepared, based on preliminary TOR established at the meeting. Administrative arrangements were finalized eight months later to mobilize a six-member project formulation team for a one-year formulation exercise, operating from headquarters in Singapore (Step 1b).

In its first month the team met key officials in the region, collected and reviewed the available documentation, and reviewed its assignment (Step 1c). It then revised its TOR concluding that 15 months, not 12, were needed to carry out the work. This issue was raised in a project meeting (Step 2a) held in the presence of a Steering Committee which consisted of representatives of participating countries and experts. The Committee agreed in principle to discuss the extra budget needed. Within a month of the project planning meeting, a project logic diagram had been prepared by the team. On the basis of this diagram, time requirements were finalized, workplan validated, and responsibilities and individual work programmes of the team members established (Step 2b).

The phase of preparation for project formulation had taken 16 months.

Discussion Topics

Points for discussion concerning the above illustrations might include the following:

- the factors which contributed to the different time schedules of Phase I for the three projects;

- the extent to which the procedures for Phase I of each project followed those described in the text and whether any departures were justified;

- the extent to which it appears that teams with appropriate multi-disciplinary skills were mobilized;

- consideration of which project got off to the better start.

PHASE II: Reconnaissance and Preliminary Project Design

Purpose and Outputs

The purpose of this phase is to 'think through the original project idea'. The team has the opportunity throughout the four Steps of this phase (3-6) to identify and review all the constraints and opportunities, the people likely to be involved, the activities which are going to take place, the technologies which may be used, and the benefits which may be generated.

A distinction should be made at this stage between projects which are specific with regard to location, species, etc. (Models A & B) and projects which are general in nature (Models C & D). The former are characteristic of almost all aquaculture projects in the private sector, and of a number of production-oriented projects in the public sector. The latter are characteristic of most public sector projects, and lack the specificities of the projects referred to above.

Nonetheless, this phase is still important in the formulation of both private and public projects. The difference in its application is one of degree. Clearly, where a project idea is already closely defined in terms of location, activity, people, technology, output, etc., the opportunities to identify further options will be limited. Even so, the phase is important as it provides the opportunity to establish that the original idea was in fact sound. Where a project idea is not closely defined, the team should use this phase to identify and analyse all ways in which project objectives can be met.

Bearing in mind therefore this distinction between the two types of projects, and remembering that even with the first type this phase presents the opportunity to ensure that the project "fits" into the physical, social, and economic environments, the phase has four main outputs. These are:

(i) An understanding of development deficiencies, potentials, and constraints of the project area/sub-sector from an overall perspective and from that of the main agents or interest groups active in the area/sub-sector.

(ii) Working assumptions about future changes in the project area/sub-sector expected to occur in the absence of any deliberate action, such as the project idea under consideration.

(iii) Identification of broad options for project design relating, for example, to definition of target groups, content of technological packages, location of physical infrastructure, and institutional participation.

(iv) Preliminary definition of a suitable project (or projects), including outline specification of project components, cost estimates, and indicative analysis of expected impact.

The work in Phase II has four Steps: (Step 3) analysis and diagnosis of the project situation from an overall perspective; (Step 4) analysis and diagnosis of the project situation from the perspective of the those concerned (the main agents); (Step 5) assessing the future without the project; and (Step 6) outline specification of a possible project. the activities of these Steps are illustrated in Figure 5, and described below. The work in Steps 3 and 4 is usually undertaken concurrently, as will be apparent from the following text.

Step 3 - Overall Analysis and Diagnosis of the Project Situation

The purpose of this Step is to understand and explain thé nature of the development problems in the area/sub-sector relative to aquaculture, and the possible solutions.

Step 3a - Review of national, regional and sectoral background

It is important that expected outputs of a projéct are considered in relation to existing and other proposed development plans and programmes for the sector, and related sectors. This usually takes the form of a general review of the broad development context in which the project is being considered. It will include the economic structure and recent trends of the aquaculture sector in the concerned area, important features of the major aquacultural systems in use, markets and price structures, institutions, organizations and management, income and welfare in aquaculture and related sub-sectors, and the government's strategic priorities and policies for aquacultural development.

Often the aquaculture sector may be small or even non-existent. If this is the case, the team should assess the resource base, and review and make analyses of directly comparable activities, such as fisheries (for marketing and price structure information), and an appropriate part of the agriculture sector (for example, small-scale farmers, and farmers engaged in pig or poultry husbandry). This comparative study would include financial and manpower allocations, farming systems and relevant social characteristics. It might also be relevant for the team to summarize briefly a situation in a similar country where the aquaculture sector has attained a level of development as to be self-sustaining and, in particular, the strategies and policies which led to this situation. If, on the other hand, there is an active aquaculture sector in the country, or the area concerned, then the team should already have a good basis of information with which to work.

Step 3b - General evaluation of the project area/sub-sector

After reviewing the broader national and sectoral context of the project idea, the formulation team then concentrates on the particular geographical area or sub-sector within which the project will probably be located. It is important here to select the most appropriate information relevant to the proposed project, and to evaluate its significance. Points of concern typically cover:

Physical features: The degree of importance to the project of climate, geology, soils, topography, water resources available for aquaculture, settlements, and existing land-use. Mapping is often the most convenient way to present this information. Attention should be given also to changes which regularly take place in

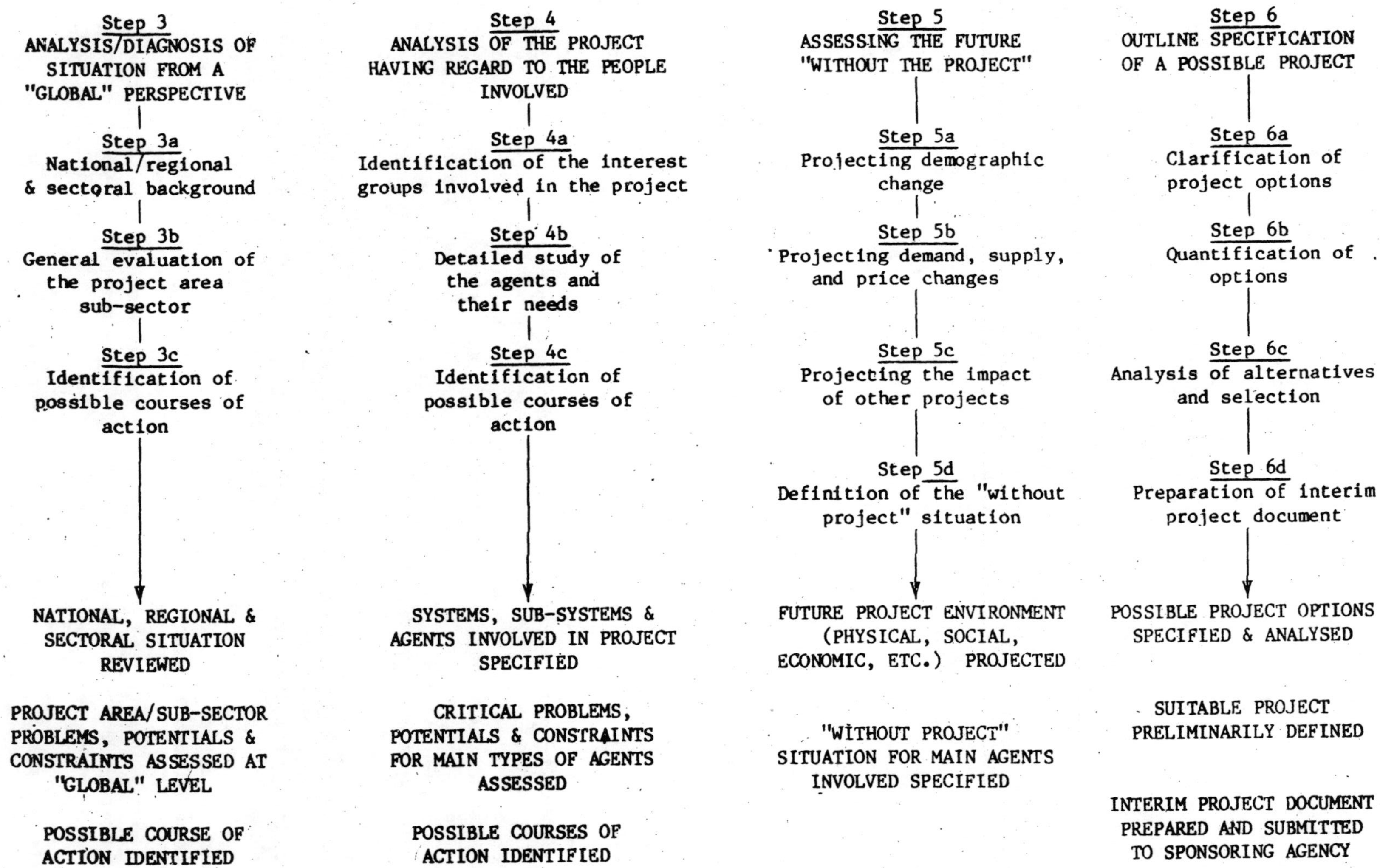

Figure 5 Phase II. Reconnaissance and preliminary project design

physical features. For example, regular changes in land use and the extent of flooding and its consequences (such as, in mapping acid sulphate soils, indicating distribution of leachates under flood conditions)

Infrastructure: The extent of and availability to the project of infrastructure, such as facilities and services, their general condition or quality, and standards of maintenance. This will also include general utilities, such as energy supplies, but if the electricity service is unreliable or fuel is often in short supply these factors should be noted as they will influence, for example, the choice of pumps for water delivery. Similarly, although at first glance there may be good transportaion links to the area or site, bridges may not withstand the weight of loaded trucks which will deliver feed to the site regularly or take out products, or the roads may be impassable at times in the rainy season.

Economic activity: Significant features of existing aquacultural practices, technologies used, levels of productivity being achieved and analyses of trends and long-term outlook, and marketing systems both upstream and downstream of the main production units. When a new technology is developed, many countries may exploit it with the result that market prices fall. Thus, in a project to produce marine fish (say) in the Mediterranean for EEC markets, the team would review the potentials in other countries with similar coastal resources, analyse the possible competition, and forecast the impact of the project on market prices.

As at other stages in project formulation, the information sought by the team at this point will be determined by the situation. For example, where the project is concerned with providing conditions to encourage production, this may only be possible if resources currently used elsewhere in the agriculture sector (land, water, labour, and capital) are diverted in some degree to aquaculture. In this case it will be important to begin establishing costs of farm inputs and farmgate prices of outputs to determine the likely financial attractiveness of fish farming to producers. If an expansion of existing aquacultural activities is planned, then economic systems which provide inputs and the marketing system will both require review and analysis to determine what changes may be needed.

Socio-economic features: These will include household characteristics, land tenure and size of holdings, water use and rights, population and migration, labour supply and employment, income levels and social welfare, and social organization. For production projects it will be necessary to determine broad social parameters related to growing marketable products, such as the availability of labour, possible involvement of women, cultural activities associated with fish and water, etc. Also at this point it may be useful to make a preliminary assessment of the rate of adoption of new technology which might be appropriate.

As aquaculture is primarily dependent on land and water, these will be mapped, competing uses will be logged and, at this point, the legal and social factors of rights to usage - which are of permanent importance in production projects - will be taken into account.

<u>Institutions and development administration</u>: Responsibilities, organization (both at local and national levels), their functions and effectiveness, and how this may affect the project should be clarified. In most countries there are government and other institutions which support different sectors, such as government departments, university departments, training institutes, research centres, etc. At this point institutions which are, or should be, supporting the aquaculture sector should be reviewed so that their strengths and weaknesses are identified. For example, where there is an agriculture extension service in place, its effectiveness needs to be assessed and, also, the feasibility of using it for aquaculture extension.

<u>Policy and strategy</u>: General aspects, which are already known from the previous phase, should not be included here but points of particular relevance to the project should be identified from discussions and visits during the field work.

If the area has no background in aquaculture then it is important to review these features in other parts of the agriculture sector. For example, there may be generous assistance from government to farmers in the form of grants and loans, and a range of support services. These should be reviewed to determine whether they would, or should, be available to the aquaculture sub-sector.

Step 3c - Identification of possible courses of action

In the preceding activities (Steps 3a and 3b) the team was concerned with setting the project idea within the overall context of the aquaculture sector. At this point this information is synthesized and analysed by the team. Using their technical knowledge and experience, the team members should be able to indicate the <u>broad courses of action</u> which could be available.

In <u>area-based projects</u> (Models A and B), the team has to consider the technological options available, together with their organizational implications and environmental constraints.

Technological choices may relate to options for systems and practices and to site selection. The team may consider, for example, a choice between the production of trout in ponds or raceways, or the cultivation of molluscs by raft or rope culture, or between farming shrimp in new or in rehabilitated ponds, or for increasing water exchange pumping through a central canal system. An organizational option in larger commercial projects may be between a central management or separate cost centres and managements for the hatchery, grow-out and marketing functions.

When considering technological systems, the team will also have to take into account the differing organizational needs of various aquaculture systems. For example, if a sub-sector has developed successfully on the basis of earthern ponds, it should not be assumed without further qualification that, because suitable water bodies are available, cage farming could be introduced without difficulty. This system requires a well developed commercial infrastructure which can provide services such as the supply of seed and feed, provision of credit, equipment, research, training and, frequently, the processing and marketing of the product. The team would need to consider in this case whether such support services were likely to be available, or could be made available.

Although the kinds of decision described above will very often have been pre-empted in the project idea, the team should examine them critically to confirm that the project idea is practical, to identify and assess other options, and to take into account potential environmental consequences of any selected option.

Key areas of environmental concern include the following:

- nutrient and organic enrichment of the water and bottom sediments as a consequence of cage farming which may result, for example, in the destruction or contamination of bivalve resources;

- partial or total ecosystem degradation which may result, for example, from the destruction of coastal mangrove swamps to construct shrimp ponds which may, in turn, reduce the availability of wild shrimp seed for both aquaculture and for recruitment into an offshore fishery;

- farm structures which may completely alter the hydrology, flora and fauna of nearby lakes, estuaries or coastal bays resulting, for example, in the damage or destruction of feeding and breeding habitats of many, perhaps also commerically important, food species;

- chemicals misused in aquaculture may threaten the health of farmed and wild populations and also of human consumers, emphasizing a need at this point in project formulation to take such possible misuse into account;

- introduced exotic species and breeds will alter and may damage local biodiversity and genetic resources.

For sector-based projects the team has to consider all possible courses of action but without going into detail. For example, in a project to increase inland fish production in the country (Model C), the team will be concerned with alternative approaches to the problem, new processes and technologies which might be introduced to achieve various quantified objectives, and also to identify existing economic activities which might be displaced. Within this approach it might consider increasing inland fish production through (i) enhancement of inland water-bodies, (ii) small-scale fish farming, and (iii) a mixture of both. Either of these activities might affect others. With the enhancement programme, which requires fisheries management, the number of fishermen might have to be controlled although those in the fishery may have a better standard of living, while a larger, managed, fishery might create more jobs indirectly than a smaller, commercialized one. With the small-scale fish farming programme, which requires inputs of land and water, it may take away these resources from local rice growers, although fish farming may provide rice growers and millers with a new outlet for their waste products as fish feed. For each option, the team would consider the boundaries of the project (i.e., which inland water-bodies and which watersheds) and make general quantified estimates of how big the project should be, with estimates of the number of fishermen and farmers, facilities required, and operational logistics. These estimates then enable the team to calculate support services required, such as number of hatcheries and nurseries, feed-production plants, ice plants, and possibly market extensions. It may also include manpower services, such as extension and training, and compare differing credit plans. It would also consider various options for the organization and management of the project.

In an institutional strengthening project (Model D) the team may be concerned, for example, with training and research in pathology and would have used Step 3a to carry out a general review of the aquaculture sector and the general constraints on it. In Step 3b the team would have concentrated, perhaps, on fish disease and adaptation of species and systems to local conditions, and on institutions carrying out research, training and extension. In Step 3c the team would consider reasons for the existing training and research centre not progressing as planned, and how relevant constraints could be overcome, and how existing institutions responsible for delivering extension services to fish farmers could be improved. The output of Step 3c, in this case, would be a series of proposals to answer these questions - which had arisen from the work done by the team in Steps 3a and 3b.

The formulation team members should discuss these options among themselves and, on an informal basis, with other concerned parties, and agree on a provisional list of the more promising courses of action which are apparent. This list will be reconsidered later in the light of information obtained when individual agents in the project area/sub-sector and their activities and needs are investigated in Step 4c.

Differences in Public/Private Sector Procedures (Step 3)

It will be noted that the work in Step 3 mostly concerns public sector projects which are general in nature (Model C), or deal with institution building (Model D)

Step 4 - Analysis of the Project Having Regard to the People Involved

Having surveyed the broad features of development in the project area or sub-sector and arrived at a first assessment of project opportunities, the formulation team should now focus attention on understanding how economic and social activities in the area or sub-sector are organized. Questions of concern will include, for example, what do people do? What do they consume and produce? What aquaculture systems are in use? Who controls water, land, credit, and other resources? Which people exercise control or influence decisions, and how?

The approach of the team to these topics is analogous to systems analysis, with people studied not in isolation but with their linkages and interdependencies on others in the "system".

In practice, many of the activities in Step 4 will overlap the field investigations conducted in Step 3.

Step 4a - Identification of the interest groups involved in the project

The provisional specification is an overall view of people who are of concern to the project, and of their activities. It is often helpful to construct a model showing their economic and social relationships. Normally this task will attempt to include all the major groups in the community who are likely to have an interest in the project.

Three main types of interest groups can be described. These groups can be called "agents", namely:

"Instrument" agents: Individuals, households, and organizations which are prime users of resources made available to the project, including not only those directly involved, such as farmers, but also others, such as suppliers and traders.

"Target" agents: Those whose material conditions and socio-economic status are of prime concern to project management, and by which the impact of the project will be mainly assessed.

"Regulator" agents: Those who control, regulate, intervene in, or otherwise exert substantial influence over the activities and welfare of instrument and target agents.

In dealing with the structure of the system, the formulation team thus considers major interests which link these various agents together. This can be illustrated by constructing flow or process diagrams.

The flow diagram of the economic and social relationships between agents in a system is illustrated in Figure 6 which shows the relationships found in a public sector shrimp-farming project (Model C).

The following agents and their links may be identified:

(a) Instrument agents, including (i) "upstream" agents, such as seed suppliers, feed suppliers (animal feed plants), village money-lenders, credit officers, equipment suppliers, private extension (consultants): (ii) producer agents, such as existing farmers and new farmers, and (iii) "downstream" agents, such as ice plant operators, shrimp processing and trading companies, and fish traders.

(b) Target agents, including farmers (who are also the producer agents referred to above) and their households, fish consumers, potential post-harvest employees (local community households), general economic beneficiaries (national households).

(c) Regulator agents, including the Department of Aquaculture in the Ministry of Fisheries.

The identification of interest groups is more difficult if no aquaculture sector exists as yet, or is small. In this case the project team will first carry out the same sort of system analysis where the proposed activity is to take place to determine how it can best be grafted on to the present socio-economic system.

Step 4b - Detailed study of the agents and their needs

From the provisional specification of the system in the previous Step the formulation team should now extend its field investigations. Field work may generate information which will cause the team to revise the system of economic and social relationships between agents, or to modify some of the criteria used to define agents (Step 4a).

The main purpose of the field work in this Step is to:

- build up knowledge of behavioural patterns of the principal agents,
- assess, where appropriate, their development needs.

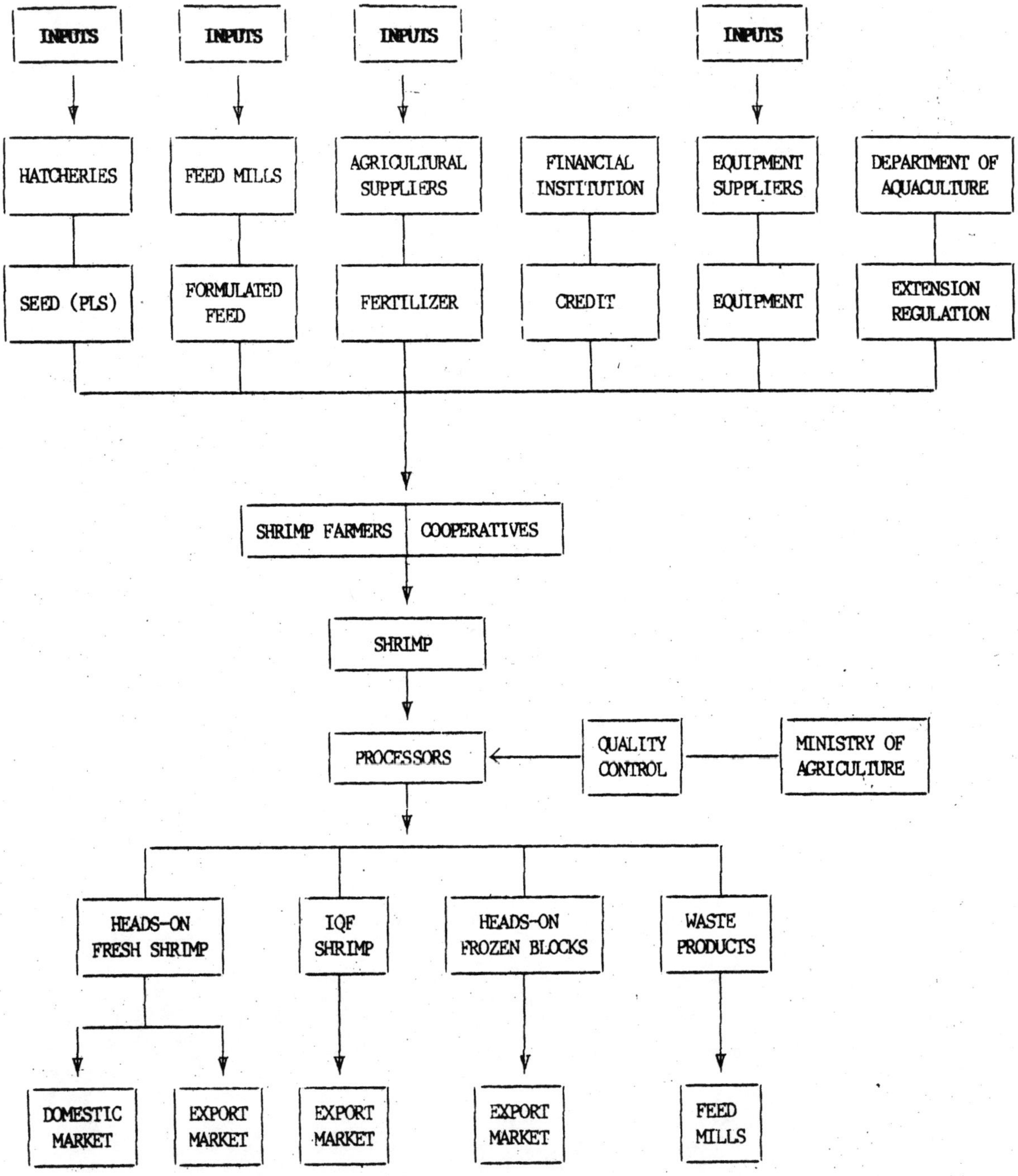

Figure 6 Flow diagram of marine shrimp farming and marketing system

Thus, the first task facing the team is gathering, in a short space of time, all the information required to enable it to achieve these two goals. Statistical surveys, based on well-prepared questionnaires, are an ideal method but take time, and may require special training of team members. Alternatively, it may be more practical to use methods of "participatory observation", that is, relatively long and detailed discussions and interviews with selected key informants, both individuals and groups, for as long a period as possible. Such meetings should include people directly concerned with the project and also those who are not concerned but have special knowledge of the area and people in it.

However, caution is necessary. In a short field trip it is difficult for a team which has no previous knowledge of an area to carry out this exercise successfully. The team leader may have assembled earlier reports, studies, books, etc., about the area (Step 1c) but, whenever possible, a sociologist from the country concerned, familiar with the area and who speaks local languages, should form part of the team. If one is not available then every effort should be made to include someone who knows the area and its people well and can guide the team. Finally, the team should not attempt this "rapid rural appraisal technique" without first referring to the extensive methodology on the subject.

In these discussions and interviews members of the team should avoid pre-conceived ideas of their own. Informants, especially the poorest members of the community, are invariably too polite or shy to contradict opinions advanced by team members. The team should always be primarily concerned with learning from all representative agents concerned, and offer suggestions only when it has a good understanding of local practices, methods, and techniques.

As an illustration of the above, shortage of credit has often been suggested by general observers as one reason why the aquaculture sector has not developed as rapidly as it might in developing countries. A project team might (say) deduce this to be true after talking with existing fish farmers who obtain credit from local money lenders. However, in discussions with credit officers in the local agricultural development banks, the team may learn that a Farmers Loan Programme (which includes fish farmers) is available with fair interest rates but fish farmers do not take advantage of it. Further investigation among farmers may reveal that they do not understand the procedural requirements, or they do not have collateral for credit (as they may not own their ponds and the value of the stock is unacceptable). Similarly, it may be found that the bank's credit officers do not understand the specific cash-flow needs of fish farmers (which differ from agriculturalists as fish are cropped over many months).

Without talking to all three types of agents the team could easily be misled into thinking that their preconceived view was correct, and there was generally a shortage of credit for fish farming. Consequently, in the next Step (Step 4c) the team would have addressed itself to this issue rather than, for example, the familiarization of bank lending officers with the credit requirements of fish farmers.

The team must make certain to include in the discussion all the potential users of natural resources which will be affected by the project. Some of these users may not be readily apparent. For example, in a project for the conversion of fish ponds into shrimp ponds, the team should take care to include those fishermen who harvest an offshore shrimp fishery which is dependent on its annual recruitment from nurseries in the mangrove swamplands, and the coastal communities whose only source of income is catching natural resources of fish seed which are then sold to fish farmers through brokers and middlemen. It is important for the team to discuss the proposed project with all such groups to make sure that it will not compete with their needs, or change their living in an undesirable way.

Again, as with Step 4a, this task is more difficult to carry out for a project where there is not an existing aquaculture sub-sector. Clearly, in such cases, agents do not have any experience from which to estimate their needs if they became involved. Consequently the team has to undertake the delicate task of educating agents, again taking care not to impose their own views, and trying to establish the response in the event of the project being established. For example, almost every production project requires the commitment of government to create appropriate institutional infrastructure. In this situation the team should spend considerable time with senior and middle-level administrators (regulator agents), advising them about typical needs, such as financial incentives (grants, loans and subsidies), and support services (possibly an extension service, hatcheries, research, education, technical training, and published information). Similarly the team should spend time talking to potential agents in the private sector, such as those noted earlier.

In summary, throughout this first task it is important for the team not merely to list individual agents' needs but to develop a critical understanding of how the needs link together within a system (in the case of an existing sub-sector), or how they would be linked (where there is not an existing aquaculture sub-sector).

The second task is to locate points within the system where major constraints are operating and the removal of which would have beneficial results. In other words, as well as understanding what is the structure and how it works, the team needs also to understand <u>why</u>, i.e., to develop a behavioural model. The team must think in a precise, systematic, and objective way about what causes things to happen in a particular project area/sub-sector. This process is only a short Step away from identifying the constraints on the system.

In approaching this activity, the team should ensure that constraints are defined as closely as possible. For example, typical limiting factors in a shrimp farming system (with reference to Figure 6) may include:

- resource deficiencies which limit yield (such as shortages of seed, poor design of ponds, seasonal shortages of water, low quality of water due to acid sulphate soils);

- institutional constraints (such as poor extension support, inadequate research, inhibiting land tenure and/or water usage rights);

- low performance technology (such as unimproved genetic material, poor quality feed, high cost of feed and/or other basic inputs, unreliable mechanical pumps, filters, and other equipment);

- economic constraints (such as lack of access to formal credit, shortage of labour, price fluctuations);

- restrictions of a social nature (such as effects on the incomes of others in the community, resistance to new ideas).

At the conclusion of this Step the team should be in a position to provide a comprehensive listing of all the system's problems, opportunities, potentials, and constraints. This provides the basis for the next Step in the formulation process.

Step 4c - Identification of possible courses of action

In Step 3c, which has the same title, the team identified possible courses of action based primarily on a general evaluation of the project area/sub-sector (Step 3b). In that task the team, subject to limitations of time and resources, surveyed aspects such as physical features, economic activity, infrastructure, socio-economic features, institutions, and government policy. In Steps 4a and 4b, on the other hand, the team had to be more concerned with reviewing the socio-economic system in which the project will operate. They addressed such questions as: which people do what? What relationships exist within the socio-economic system in the project area/sub-sector? What are the limiting factors which prevent the system from performing better?

This Step can best be described as consisting of four activities. They are:

- revision of the possible interventions identified in Step 3c in the light of the work done in Steps 4a and 4b;

- final identification of the different groups of agents to be involved;

- identification of complimentary actions within the proposed project which are required elsewhere in the socio-ecomomic system as a result of the primary intervention(s);

- discussion of proposed courses of action.

First, in this Step the team synthesizes the work done in Step 3 - which may be typified as being concerned with "things" - with the work done so far in Step 4, which has been concerned with "people". This is done by "fitting" the "people analysis", of Steps 4a and 4b, to the possible interventions proposed in Step 3c. Thus it is possible that a course of action identified in Step 3c might have to be dropped, perhaps because it is now apparent that it would be inimical to the existing social and economic relationships between agents, or perhaps because the analysis in Step 4c has identified needs of agents and constraints upon them which give other possible interventions a higher priority.

At this point the team should become aware that the project may not bring about the changes intended, or may bring about other undesired changes. For example, a project for producing fish in net-pens or floating cages in large inland water-bodies may physically obstruct traditional small-scale fishermen and deprive them of their living. Similarly, in the above illustration of the shrimp project, the team may decide that converting all existing fish ponds and fish farmers into shrimp ponds and shrimp farmers would have serious repercussions on the economic and social conditions of the traditional fish seed collectors, and there would be loss of fish to local markets. In this situation it may conclude that it will be desirable that only half the fish ponds available should be converted to shrimp production. Similarly, it may have found that destroying all the mangrove swamplands to make new fish ponds may destroy the offshore fishery on which many fishermen depend, thus new ponds should be constructed just behind the peripheral edge of the mangroves, so preserving natural nursery grounds.

Second, after these adjustments have been made, the team should now be in a position to consider who are to be selected as main "target" agents or beneficiaries, in the case of the shrimp project, or who constitute the active receivers or users of projects designed to improve technology, or receive infrastructure, or a line of credit, etc., depending on what is to be introduced by the project.

At this stage the best the team can do, on the basis of the limited field work it has carried out, is to make a first estimate of the number of target agents. The adoption rate will be influenced by such factors as the attitudes of target agents to new technology, the availability of inputs, the availability of credit, and an estimate of the number of agents prepared to make the commitment of time required.

Third, clearly because intervention in one part of the socio-economic system is likely to have consequences elsewhere within it, the team may be required to identify additional courses of action. For example, intervention in the form of private sector investment in shrimp production may require new investments in ice plants and a cold storage facility for the shrimp, destined for exports, to maintain quality control, and possibly a small unit in the Department of Public Health to monitor it; or a project to encourage pond farming may result in a need by farmers for some form of cooperative or association which, initially, might receive government support. These additional interventions elsewhere in the system will form new components of projects.

The need for such institutional changes, as a consequence of the introduction, say, of the production and marketing components of a project, may be immediately apparent, through the need for extension services, demonstration farms, study tours, etc. In contrast some required changes may not be so apparent, and the team therefore should have a good understanding of the responsibilities and functions of institutions, particularly in government. For example, if a team is poorly informed, it may not realise that where a project is concerned with the introduction of non-indigenous species of fish into the country, or the movement of adults, eggs, and juveniles, then the government should have a policy, complete with regulations, towards introductions and transfers, and possibly a small government veterinary unit, complete with quarantine station, to authorize and monitor such activities.

Fourth, the members of the formulation team should discuss these ideas among themselves and rank them in order of importance and in terms of their impact on agents. They should then discuss their preliminary conclusions with appropriate parties outside the team who are able to comment and give advice.

Differences in Public/Private Sector Procedures (Step 4)

In private sector production projects (Model A), the three tasks of Step 4 may be omitted altogether, or they may be included when useful to the project sponsor. For example, the project sponsor may be concerned that his production project (which, say, requires low-cost labour) fits into the local socio-economic system and that no constraints will be imposed on the project. Sometimes this Step will be fully developed by the team or, on other occasions, the project sponsor will be fully aware of the socio-economic structure in the project area; for example, the international corporation converting its sugarcane lands to catfish production.

In public sector projects (Models B, C, & D), however, it will be clear that all three tasks in the Step are usually very important. Where it may not be so important is in projects which are analagous to private sector projects, such as a state-owned farm.

Step 5 - Assessment of the Future "Without the Project"

The team should now have a comprehensive understanding of the current development situation in the project area/sub-sector. It will have ideas not only of what could be done to improve it, but also on what could happen either independently of the project (whether it is implemented or not), or in the absence of the project. These prospective changes which will take place if there is no intervention with a project are now analysed and evaluated so that the impact of the project can be assessed.

Step 5a - Projection of demographic change

The purpose of this activity is to consider how the population in the project area, or in areas relevant to the main activities of the project, is likely to grow and change over the life of the project, paying careful attention to complicating factors, such as migration and employment.

Demographic changes are particularly important in the planning of aquaculture projects because of the finite nature of land and water resources, the two principal needs of aquaculture. For example, a significant increase in population in the project area may result in increased competition for relatively scarce freshwater supplies. On the other hand a project may induce such changes. If a new area is developed, such as mangrove swamplands described in the illustration, greater access to coastal beaches may be offered which may encourage tourism and subsequently hotel development.

Step 5b - Projection of demand, supply and price changes

A fundamental requirement at this stage is to assess the level of demand for the outputs of the project. This assessment requires clear distinction between the need for a commodity or service (that is, a requirement determined by non-market criteria, such as minimum food energy intake), and the demand for that commodity or service (that is, the quantity or level which consumers are willing and able to acquire at the particular price offered).

There is an established methodology for assessing demand. The main determinants of demand are population growth and changes in per caput income. If anticipated changes in demand have not been estimated in a sector study, the formulation team will usually be able to obtain forecasts of population growth - in larger countries on a regional basis - from the central planners. The extent to which changes in income will affect the demand for fish is often a complex issue. Very often, however, the application of a broad estimate of income elasticity of demand - the amount by which demand will change in response to a given change in income (which may also be obtainable from the central planners), will provide a crude but usually satisfactory measure of the likely magnitude of demand. In those areas where there is a combination of low incomes and a high reliance on fish in the diet, the income elasticity of demand can be quite high. For example, for every 10% increase in family income, demand for fish may increase by 5%, assuming no increase in the price of fish relative to other goods.

Projections of supply will be dependent on the particular circumstances of the project. In some cases it will be appropriate to take into consideration the supply of fish from all sources and in others only from aquaculture. Other factors may include types/species of fish, fish products, and consumer attitudes.

Once demand and supply projections have been made, it is necessary to bring them into balance. In many instances it will be found that there is a widening gap between demand (fuelled in many of the poorer countries by increasing population rather than incomes) and supply. Where this occurs, the real price of fish (that is, its price relative to other goods) will rise.

A comparison between projections of demand and supply in the absence of the project results in price projections in a without-the-project situation. These enable the formulation team to assess the future size of markets for outputs of the project and to indicate the directions and magnitudes of change in market prices with and without the project.

In projects which have the objective of producing product for export, for example marine shrimp in the illustration given above, projections of demand, supply, and price changes are important elements in determining the financial feasibility of the project. However, it is very unlikely that the incremental supply of shrimps by the project will influence world supply and price levels. However, where a project is to supply fish to a relatively small local market, it will be important to determine whether that market could absorb all the project's output without a fall in price.

Step 5c - Projection of the impact of other projects

In this task it is important that the team is aware of other projects (ongoing or planned) which may have a possible impact on the project which is being formulated. This is particularly necessary for any projects which compete for land and water resources. Also not to be overlooked are projects which absorb human resources and skills in the counterpart institutions which might be associated with the project. Experience shows that many governments often overcommit themselves to development projects without taking on additional staff, or releasing staff from other responsibilities.

Taking other projects into account usually presents little difficulty for the team when they are already ongoing, but it is often more difficult to learn of pipeline and proposed projects. Furthermore, it is often uncertain whether such projects will ever be implemented fully, in part, or after some delay. In such cases there may be no other option but to postulate alternatives of the without-project situation, either with or without one or more of those other ongoing or potential projects.

Step 5d - Definition of the "without project" situation

It is implicitly assumed that any project will result in additional outputs from the project area/sub-sector concerned, that is, a greater quantity and/or better quality of goods and services will be produced than would otherwise be forthcoming. Thus a clear and precise assumption of what is "otherwise forthcoming" is particularly important.

For this task the team must avoid the temptation to jump quickly to the convenient but often erroneous conclusion that the situation will remain much as it is in the absence of the project. Moreover, it should consider that if the project under formulation is not implemented another project might eventually be implemented in its place. The effect of the other project is also part of the "without project" situation.

Consequently, it is important at this point to produce a practical statement which represents the team's best forecast as to what may happen in the project area/sub-sector if the project under formulation is not implemented. This forecast serves as the base line for evaluating project results. For example, if there are no foreign earnings from shrimp exports in the next five years without the project, then all earnings from exports as a result of the project (say, US$ 20 million) can be claimed as a project benefit. However, without the project fish production in the area may remain at its current level of 5,000 tonnes per annum. With the project this would be halved. Similarly, these kind of calculations may be made for farmers' incomes and employment figures, etc., and presented as "results".

Differences in Public/Private Sector Procedures (Step 5)

All four tasks within this Step are common to the four aquaculture project Models described earlier. Clearly, the tasks are likely to be more complex in multi-component public sector projects (Models C) concerning many agents in different groups.

Step 6 - Outline Specification of a Possible Project

While in the previous Steps there has been a relatively close parallel between the formulation procedure normally adopted for both public sector and private sector projects, in Step 6 there is, in almost every case, a wide divergence in approach.

In private sector project formulation quite definite answers will have emerged to the following questions during the work in Phase I (Steps 1 and 2) and the early parts of Phase II (Steps 3 and 4) about the project:

- Where should the project be located?
- Who should participate in the project?

- What should the project be doing?
- How big should the project be?
- When should the project start?
- Which organization should manage the project?

In these projects the degree of detail in the outline specification is considerable, and therefore the cost estimations which form part of this Step are in similar detail. The end product, the interim project document, is therefore the prefeasibility study report.

In public sector project formulation the team has not yet been able to answer the questions listed above. It is still at the stage of sifting and sorting the different courses of action which will achieve the project objectives, and which were identified in Steps 3c and 4c.

In Step 6, therefore, the team will:

- clarify all possible options;

- identify which options appear to be superior to others;

- present the selected options in the form of a specific project proposal.

The end product, as with private sector projects, is the interim project document which, reflecting the formulation process in public sector projects, is frequently called the identification report, or the reconnaissance report.

Step 6a - Clarification of project options

As noted above, this Step is undertaken in public sector projects, and not usually by those in the private sector.

The six questions posed above nave now to be given fairly definite answers at this point. These should result, collectively, in the outline specification of the project options. The sequence of the questions is not significant, as all answers must be considered in parallel, but different combinations of answers collectively form different possible project outlines.

The options have to be considered by the team within the context of the project objectives (Step 1), the criteria under which the project is being formulated (Step 2), the project area (Step 3), and target agents (Step 4). Factors which will be taken into account when answering each of the questions may include the following:

(i) Location of the project. The larger boundaries of production-related projects are usually defined geographically in the context of activities to be carried out, availability of inputs, and possibly in terms of institutional responsibilities. For example, in a project to develop culture-based fisheries, the team will compare locations having regard to such factors as the sizes of different water-bodies and potentials to support fisheries, existing and potential future fishing effort, siting and operation of hatcheries, prospects for introducing effective management practices, availability of fishery inputs (boats, nets, engines, fuel), and environmental impact. Or, in a project to enhance national capabilities in national research and training, the team will compare locations for such factors as existing institutional strength, proximity to the sub-sector, policy issues (at national, state, and local levels) and physical facilities (research laboratories, information resources, accessibility, accommodation).

(ii) Participants in the project. There may be options concerning who should participate in the project. For example, in a project to increase mollusc production, the team may compare options for new producers in new areas, assisting existing producers to expand in established areas, or a combination of both. Alternatively, the team may consider that increased production could be achieved by reducing waste in the transportation and marketing networks, and "downstream" agents must be involved.

(iii) Activities of the project. For production-oriented projects, the options may be few, as the production systems and practices will be determined, to a very large extent, by external factors. While the options may be limited in number, they do nonetheless exist. For example, to increase mollusc production, the first consideration may relate to the choice of species, namely oysters or mussels, or both. With regard to technological options to increase yield, the options may be to produce them on floating or fixed structures in the sea, or to improve management of any existing beds, or both. Similarly, attention may be given to the post-harvest stages, such as the introduction of depuration and processing to prolong shelf-life. In some production projects, say to increase fish production in certain areas of developing country, there may be no alternatives to simple pond culture other than increasing levels of intensity. Other projects have more easily identified options. For example, in a project to introduce new hatchery technology, alternatives may be to train hatchery managers in national training courses, or on-the-job training courses; in a project to advance education options may relate to introducing graduate courses at national universities, or taking advantage of graduate programmes overseas.

(iv) The size of the project. The maximum size of the project may be constrained by external factors, such as a national plan for development, or internal factors, such as total allowable cost or designated area. For production-oriented projects, options will be quantified production goals. Factors which will influence these options may include constraints referred to above (location of the project and participants), and perceived risk. Thus, options may be developed relating to the size of the project and, furthermore, there may be options relating to its phasing. For example, where there is perceived to be a relatively high level of risk the formulation team may array as options (a) a relatively large project without a pilot stage, (b) a pilot stage within a larger project, or (c) a pilot project only. For other types of projects, such as institutional strengthening, options may be severely limited by predetermined numbers of graduates required, or by institutional resources available. Options may be increased, however, when, in the absence of any sector development plan, the formulation team proposes different scenarios for institutional building based on a range of estimates of production by an aquaculture industry.

(v) Coordination and scheduling of project activities. Once again, under almost all circumstances, the options for production-oriented projects are limited by the reproductive life-cycles of the species concerned. Thus it is easy for a project schedule to slip a year if an activity which anticipates a breeding season, such as hatchery construction, is delayed. In a project to enhance production through culture-based fisheries, where it is apparent at this stage that a substantial amount of time may be lost due to unavoidable delays in construction of hatcheries and other facilities, an option may be to consider purchase of juveniles elsewhere and implementation of improved management practices prior to hatchery operation. Available options are also likely to be limited in other projects, such as institution building. In research and training projects, for example, principal options may concern specific timing of activities under the project, which may have cost implications.

(vi) Organization of the project. In small aquaculture projects there are few if any options for organization and the project must fit within the existing government structure. In large projects identification of the right management institution is often a sensitive question, as the most appropriate may not have participated in project conceptualization at all. This is particularly true of inland aquaculture projects which may have been developed through (say) the Ministry or Department of Fisheries but, because of their potential involvement with water resources, farming cooperatives, and interior extension services, are best placed within the Ministry of Agriculture. Alternatively, if the agriculture sector is already large and dominant in the country, it may be preferable to use an existing parastatal body, such as a regional development authority, to manage the aquaculture project to give it some independence and a specific identity.

Step 6b - Quantification of options

In private sector projects the team will have omitted Step 6a and proceeded directly to this task. Here the team will quantify quite accurately the main technical and biological criteria used to define the physical design of the commercial facility. By comparison, in public sector projects the team having only established a number of possible options, will now quantify them to a degree by that a judgement can be made about them later. The purpose of this task for these types of projects is therefore to make preliminary estimates of the costs and benefits of each one as a prerequisite to making that judgement.

Private sector projects

In private sector projects quantification of the technical and biological criteria is called bioprogramming. It involves assembly of all the design features, and quantification of all components which describe functions and operations. Bioprogramming can be simple or detailed, depending on the nature or purpose of the project. Its elements are as follows:

(a) Biological design criteria. These identify the basic environmental requirements to be fulfilled in the design of the facility for the species to be cultured, and the optimum conditions.

Careful research of facts is necessary, usually with reference to research studies, reports of other facilities, and trends in culture techniques. The key biological criteria include species description, behavioural characteristics, culture techniques, environmental requirements (in terms of density, water exchange, water quality), nutritional requirements, growth rates, mortality rates, diseases, etc.

(b) Facility characterization. This activity integrates the biological design criteria and production goals with the proposed site of the project and transposes them into a functional physical and operational plan. The characterization of the facility is based on production schedules (taking into account all stages of development as well as at harvest) and the number of units required. Several iterations may be necessary until a realistic project is developed in terms of biological, engineering, and financial constraints. For example, the site may be limited in size, and therefore with standard engineering design its production capacity is limited and possibly not profitable. If a higher technical level of engineering is employed, at additional cost and risk, increased production is possible through intensification, and increased profit. Alternatively, the site may not be limited in size, in which case it is necessary to develop sufficient of the site to meet production and profits for the least cost in construction and future operation. Thus the task is to characterize facilities which optimize all these factors.

(c) Programming infrastructure. This element deals with specifications of the infrastructure, in particular the water system for the site and all production units, and all production and support spaces. For water requirements, a water process diagram is developed which indicates the volume of each type of water required by each production unit and support space, and any seasonal variations. Production spaces will include the number, size, and other features of all grow-out spaces, and the hatchery, nursery, brood ponds, spawn tanks, quarantine tanks, etc. Support spaces will include dimensions and services required of all features, such as research laboratories, offices, storage rooms, feed stores, workshops, freezer equipment, boat ramps, etc.

(d) Site studies. This activity identifies the characteristics of the site. Primary site criteria include land availability, land ownership, and cost, as well as its topography, water availability, water quality and temperatures, water rights, and other competitors for the water. The other primary criteria are the qualities of the soil (chemistry, permeability, and compaction properties). Other features of the site are also important, such as proximity to supporting infrastructure (towns, utilities, markets), and access (roads and bridges).

The site can also be described by its local political and socio-economic factors (Step 4b), the climate and its potential for major disturbances (earthquakes, floods), and adjacent land uses (Step 3b). It is also necessary to be aware of any leases or permits required for developing the site and use of water, and other environmental constraints which might be impacted by the project. Finally, it is useful to know if the site has natural predators of the species to be cultured, or if there are other pathogenic organisms in numbers to add further risk to the project.

(e) Schematic design. This element translates the criteria into a workable design. It identifies, assesses, and recommends the design which meets production goals optimizing the features of the site. The schematic design assembles information which:

- describes the facility design concept selected;

- a site lay-out, showing location of all major components on the site and reflects topography and operational relationships;

- floor plans for all major support buildings;

- a schematic diagram of the hydraulics of all water requirements for each component complete with elevations;

- a topographic survey.

Sketches are prepared to illustrate the concepts. This information is then used for estimating preliminary construction and operating costs, and developing a project construction schedule.

Two further elements are then included:

(f) Financial analysis. This element provides a preliminary financial analysis of the proposed project, based on the objectives and the cost/benefit requirements of the project sponsor. Reference here is also made to the original market assessment and marketing strategy for the products. Invariably the analysis is made in the form of financial spread-sheets, complete with calculations for internal rate of return.

(g) Environmental analysis. The last element considers the relationship of the project to the environment, and considers both positive and negative impacts which may follow. In some cases the team may be required to research national laws and local regulations regarding the environment, to make sure that the project does not conflict.

This Step concludes with the production of a prefeasibility study report, which is a compilation of the elements (e), (f), and (g) above. This report is complete in as much as the project sponsor can agree to financing, or can use it to secure financing. In addition, the document is an accurate record of the project, which may become a guide for all subsequent stages of construction and operation.

It should be noted, however, that during schematic design (e) above, the team may conclude that there is more than one way to lay out the site. There may be simple options for design but with alternative costs and risks. For example, a pumphouse could be located on the coast for maximum efficiency but this would require costly coastal shoreline construction to resist wave forces and lengthy transmission lines for electricity and large pipelines for water. Alternatively the pumphouse could be located near other facilities inland where it would not be as efficient but there would be no electrical transmission lines and tidal water would be brought in cheaply through a simple canal. It may also be that large ponds would be cheaper to construct and operate but when harvested the volume of fresh product from each could exceed the projected market demand at any one time. Alternatively, smaller ponds could be constructed at greater costs, but future management and harvesting might be better geared to local market patterns.

When such options arise, as they invariably do, the project formulation team will then follow, but to a lesser degree, the process of dealing with options which, as noted in Step 6a, are typical of public sector projects.

Public sector projects

The quantification of options in public sector projects which have components of production deals in effect with these same elements of bioprogramming and, where possible, uses the same criteria. It is unlikely to be as detailed, however, as the information required can be restricted to that permitting the analysis of each option in Step 6c before one is chosen for further development.

This can be demonstrated by reference to the illustration, used in Steps 3-5 to (Figure 6), which deals with a project to produce marine shrimp to earn foreign exchange by constructing new ponds in coastal mangrove swamplands and/or converting existing fish ponds. In Steps 3c and 4c the team had identified possible courses of action, namely options for constructing new ponds in the mangrove swamplands, converting existing fish ponds into shrimp ponds by making them deeper and increasing water flow, or a mixture of both. The team should undertake bioprogramming for each option as if it were an independent private-sector project. In other words, it will assemble:

- biological design criteria,
- facility characterization,
- siting, and
- schematic design.

As with the private sector projects, the team should also estimate capital costs and operating costs but a lower level of accuracy is acceptable. Information on yields and prices will enable estimates to be made of revenues and foreign exchange earnings for each option. Profits can then be calculated on a year-by-year basis, and a preliminary financial internal rate of return (FIRR) be estimated.

Preliminary estimates can then be made of net economic and social benefits. In calculating the economic internal rate of return (EIRR) there might be taken into account, for example, the number of new jobs created and incremental increases in incomes generated by the project. A preliminary estimate might also be made of economic costs, for example, shadow price of labour and opportunity cost of foregone fish production. Judgements can also be made about other benefits and costs, such as diversification of economic activity, creation of self-reliance and management skills, and even transfer of skills and technology. Similarly, qualitative judgements may be possible with regard to other costs and benefits, such as possible reduction of the shrimp fishery, or loss of biological diversity through destruction of mangroves.

The analysis of the three options, in the light of all this information, is then carried out in Step 6c.

Step 6c - Analysis of alternatives and selection

Whether the project is in the private or the public sector, each quantification and qualification derived above constitutes, in outline, a different project. In principle each option may be feasible, but at this point the team must decide which is the best.

The basic reasoning behind each design option is reviewed once more to determine if it conforms with the objectives of the project. Once this is done each design is submitted to performance tests concerned with four particular criteria, namely:

- Financial criteria. Money is the unit of account. For example, what will be the project cost? How will it be financed? Are the financial incentives likely to induce the responses required?

- Economic criteria. Financial aspects concerning individual agents are integrated in resource utilization as a whole. For example, would the resources invested in the project be better utilized there than elsewhere?

- Social criteria. These account for the distribution of benefits among various groups of society. Would the project contribute to achievement of the country's social policy objectives?

- Environmental criteria. The need to exploit natural resources must be balanced against its effects and impact in the future.

In public sector projects, providing answers to the many questions appropriate within the four criteria is one of the most difficult tasks of the team. There is no obvious answer to weigh the performance of one design option against another. The team therefore has to use its judgement. One useful tool to assist the process is the decision matrix. This is a two-way table by which the different project designs considered are entered against specific objectives and design criteria, forming cells where the performance of the corresponding combination is noted.

The way in which the matrix is interpreted depends on the type of project and its objectives. If, for example, a country's sectoral objectives include the promotion of living standards and employment in rural areas as a means of slowing urban drift, the number of householders benefiting from the project is the important criterion; in a project to earn foreign exchange, financial and economic criteria are more important.

All projects carry a degree of risk, some more than others. The team has also to make a judgement on the level of risk for each option proposed.

Table 2 shows a decision matrix for the shrimp-farming project developing new ponds in coastal mangroves, and converting existing fish ponds into shrimp ponds by making them deeper and increasing water flow. The matrix lists the criteria most appropriate to the project. This table shows the way that it might have been constructed by project formulation team.

Table 2

Project decision matrix
(coastal shrimp farming project)

Option A: Construction of new shrimp ponds.

Option B: Conversion of existing fish ponds into shrimp ponds.

Option C: Combination of new ponds and existing ponds for shrimp production.

Objective/Design Criteria	Option A	Option B	Option C
ECONOMIC			
EIRR (% pa)	27	24	25
Project exchange rate (local currency)	5	4	4
FINANCIAL			
FIRR (% pa)	12	20	17
Capital costs to Government (US $ mill)	3	1	2
Government recovery of costs (%)	85	85	85
External capital assistance	Yes	Yes	Yes
SOCIAL			
Employment (Management, clerical, & labour)			
Hatchery	16	16	16
Ice plant	0	30	10
Construction (man/years)	3 000	500	2 000
New self-employment householders	250	0	100
Households with enhanced income	250	600	400
Geographic dispersion of activities	0	++	+
Self-reliance in economic management	+++	+	++
Transfer of skills and technology	+++	++	++
Market effects (local fish consumption)	0	---	-
ENVIRONMENTAL IMPACT			
Destruction of ecosystem	---	0	--
Possible affect on fisheries	---	0	--
Possible physical/chemical effects	--	0	-
RISK			
Positive level of risk	+++	+	++

Scale: 0 = neutral: - = negative factor (low to high): + = positive factor (low to high).

As the project was finally formulated, there were three options. These are summarized below.

Option A: Construction of 2 500 ha of new shrimp ponds in reclaimed coastal mangrove swamplands with an annual yield of 10 000 t in two cycles, together with a hatchery. The major part (80%) of the capital required for construction for the ponds originates from external assistance with the balance povided by the national government; the balance of the capital requirements for financing construction of a hatchery, and technical assistance is also provided by external assistance. The financial support from the donor is provided through a line of credit to the Agriculture Development Bank for on-lending to 250 individual farmers, and to the parastatal corporation which will build and operate the hatchery. New farmers will be recruited from the households of traditional fish farmers in the area.

Option B: Conversion of 4 000 ha of existing fish ponds to shrimp ponds with an annual yield of 10 000 t in two cycles, together with a hatchery. The capital arrangements are as before but, as the fish ponds are widely dispersed (among 600 farms), there is an additional credit component for three ice plants.

Option C: A mixture of the above, with 1 000 ha of new ponds (100 farmers from existing fish farming households) and 2 000 ha of existing fish ponds (300 farmers) with combined production of 10 000 t. Capital arrangements are as before but only one new ice plant is required.

An interpretation of the project decision matrix (Table 2)

Economic: All options have high and similar EIRRs, with the highest in Option B because of the higher self-employment benefits generated by the project. Option A generates economic benefits through a higher employment in construction. All require some domestic capital, which is estimated to to generate US$ 1 for every 4-5 units of local currency. Note, however, that these EIRRs are subject to further qualification in the Environmental Impact Analysis (see below).

Financial: Option B has a rather higher FIRR than either of the other two projects because of the lower construction costs.

Social: Option B has higher benefits than the other two options because of the higher level of participation by self-employed householders (600) in the project. Employment in the ice plants under Options B and C would be seasonal, and employment in construction in Option A will occur only once. These may be disadvantages in attracting labour, and some problems may arise if labour is imported. Overall, Option B provides the greater flow of benefits, as well as upgrading self-reliance in the local community by introducing new skills into it. Option A provides the lowest social benefits. Option B, and to a lesser extent Option C, results in the loss of fish to the local markets. The source of new farmers for Option A (250) and Option B (100) is assumed to be from among traditional fishpond households. The investigations carried out by the team (Step 4b) have indicated that (a) here should be little difficulty in recruiting young farmers with the necessary aptitudes and skills, and (b) their recruitment should have no adverse affect on the existing traditional fish farming activity.

Environmental: Option A has the greatest environmental impact as it requires the greatest destruction of coastal mangroves. In addition to this loss of a coastal ecosystem lands, the loss of the nursery grounds for young shrimp may affect both the artisanal and industrial shrimp fisheries. This would have a significant impact on the EIRR of Option A, and to a lesser extent in Option C. Options A and C might enhance local productivity by the discharge of nutrient-laden water from the new farm complexes, but Option A, because of the size and associated greater use of water, may result in scouring and coastal erosion.

Risk: There is a risk for all three Options associated with the "take-up rate" of the new technology. In Option B, however, it would be expected that this risk is lower as the existing fish farmers can use the technology in some of their traditional activities. Both Options A and C might be considered to have high risk associated with the construction of large new areas of shrimp ponds because it adds significantly to the complexity of the project.

Conclusion: All three Options have similar EIRRs and FIRRs, and the government recovery of costs is estimated to be similar. Militating against the selections of Options A and C would be the high level of risk associated with them, compared with Option B; a significantly higher environmental impact, and possibly greater adverse social repercussions which might accompany the employment of a large labour force for construction. In favour of Option B is that it has the largest number of beneficiaries (implying the largest distribution of benefits). In favour of Options A and C would be the generation of new self-employment, and larger benefits in the form of increased self-reliance in economic management, and increase in the transfer of skills and technology. Option A would have no consequent loss of fish to local markets, Option C would incur some loss, and in Option B the loss would be significant.

After balancing the arguments in favour and against each of the Options, the team would probably select Option B on technical grounds, as the ponds are already in place and operated by traditional fish pond farmers. It has a low level of risk so far as cost over-runs and low "take-up rate" are concerned; neutral environmental impact, and avoids, to a very large extent, possible adverse social repercussions during construction.

However, the government and the external assistance donor, when reviewing the options, might prefer Option C on the grounds that the loss of fish to local markets is reduced, while only slightly fewer households (450 compared with 600) are beneficially affected.

Step 6d - Preparation of interim project document

The outline specification of the formulation team's preferred project design (or designs) is now presented in proper form to the project sponsor.

Referring back to the beginning of Step 6, it will be recalled that the end product, the interim project document, was called the "prefeasibility study report" in the private sector, and the "identification report" or reconnaissance report" for projects in the public sector. In the discussion of this Step, attention has been directed to the differences in approach and level of work undertaken between the two sectors. Clearly, interim reports reflect these differences in their contents.

A prefeasibility report would describe a properly evaluated and functional facility with the projections of costs, and construction and operations schedule, made in sufficient detail. In relatively small or simple projects, where the overall cost is, say, up to US$ 500 000 (in 1990 prices), a decision might be made by the project sponsor to proceed immediately to implementation. In larger and more complex projects there would almost always be a requirement on the part of the project sponsor for the formulation process to proceed to the Project Design Stage (Phase III).

The identification report or reconnaissance report usually takes the form of a clear description of the project's nature, structure, dimensions, and impact with all the choices concerning project variables being explained and, if necessary, justified. The lay-out of the report might be described in terms of the reverse of the procedure which has been followed by the team through Steps 3-6, described above, that is the report would highlight for the reader why the project is the best, and for what reasons.

At times, as noted above in Step 6c, there may be more than one possible project design. The team may also conclude that either more detailed design work is required (Phase III) or, simply, no project properly satisfies the TOR. Provided that the team has done its work correctly and professionally no failure is associated with such a conclusion.

Each report ends, whatever its evaluation, with clear recommendations whether to proceed, or not. If the former, specifications and TOR of the next phase are given.

> **Differences in Public/Private Sector Procedures (Step 6)**
>
> The four tasks of this Step are quite different for private sector projects and public sector projects. The private sector project (Model A) is more concerned with accurate quantification (Step 6b) and financial analysis (Step 6c). In public sector projects (Models B,C, & D) the emphasis is usually on the evaluation of all the alternatives identified, and makes use of all Steps. While this includes quantification and financial analysis where possible, it is not usually carried out to the same degree in private sector project formulation.

Illustrations of Phase II: Reconnaissance and Preliminary Project Design (Steps 3-6)

Project 1 - A Commercial Fish Hatchery Project in a Mediterranean Country

Following preparation for project formulation (Phase II), the investor returned to the country and signed contracts with local companies for a topographical survey of the site, soil analysis, and water quality analysis respectively. A three-man team from the consultants arrived a month later, which included a hatchery engineer (team leader), a fish culture biologist, and an economist.

The team spent the first week together reviewing the national and regional background of the project (Step 3a) as it related to marine fish farming in the Mediterranean, its projected growth, and need for seed resources, and made a general evaluation (Step 3b). They brought with them much relevant information on the subject and from all information developed a general course of action (Step 3c). They spent time with the investor confirming his needs, and the economist then carried out a market survey to determine the national market for seed and that in the six other countries in the Mediterranean region which were both potential markets and sources of competition (Steps 4a and 4b). His separate report included a market study (Step 5b), a review of the development situation in other countries (Step 5c), a socio-economic review including a study of local labour costs, etc. (Step 4c), and confirmed the general course of action (Step 4c). The investor himself provided all information regarding national and local legislation for aquaculture projects, and current government incentives and support services (Step 6b).

The engineer and biologist spent their remaining three weeks quantifying the project (Step 6b). The biologist produced all biological design criteria for sea bass and sea bream species and, with the engineer, developed the programming infrastructure, and translated this into the facility specification. The topographical survey and water quality analysis were received in the second week, but the soil analysis was only received just before the team left to return home. The team had discussed the project with the investor to summarize their progress before leaving.

In their own offices the team completed the bioprogramming document (Step 6b). The economist's marketing report was a separate attachment. In laying out the schematic design for the site the team identified two options. The nurseries could be located near to the hatchery which was on a flat but elevated area; alternatively, they could be located at sea level to avoid the need to pump large volumes of water uphill but this option would require more of the lower area to be blasted flat to accommodate the nursery tanks. There were also two options for the marketing of seed. The economist reported that a number of farms in the region were anxious to secure regular supplies of seed under contract, with annually negotiated prices, but with penalty clauses for default. The alternative was to sell as available on the open market. His report also stated that two similar hatcheries were being built in France and in Italy (Step 5c), and that the combined production of some 20 million juveniles per annum would be far more than was required in the region for the next ten years.

At their office, the team completed financial analyses for four options (two site lay-outs, and production of five million and three million juveniles <u>per annum</u>), and a brief environmental impact statement to cover the required blasting and discharge of enriched water from the site back into the sea. The team analysed the options (Step 6c) and recommended low production figures of three million juveniles (with service installations built for expansion) to be sold under contract to farmers in the region, and the nursery located at the lower level to avoid high costs of pumping water uphill. The FIRR was marginal until the facility was expanded.

The team leader then returned to the country and presented the prefeasibility study report to the investor, and explained the recommendations.

The procedure had taken eight months.

Project 2 - A Brackishwater Shrimp Farming Project in a Southeast Asian Country

This area-based project concerned the construction and/or rehabilitation of canals and ponds to achieve higher yields of shrimp, and provision of a number of government support services. The TOR required the team to consider a number of areas in two regions of the country and to rank proposals in order of priority (Phase II of the formulation exercise), and to plan to feasibility study level a project for immediate implementation (Phase III).

The review of the national and sectoral background (Step 3a) examined the government's strategy and policies with regard to pond farmers and provision of support services. As a result of previous visits to the country by team members, the government position on a number of these issues was already known. It was confirmed that the government's fisheries development strategy included a large measure of support to pond farmers to bring about an increased production of shrimp through aquaculture. It was also confirmed that this new source of shrimp should be processed, whenever possible, at existing shrimp processing plants. The team visited an existing shrimp research and extension centre, and was shown the different extension methodologies employed.

The team began the major task of general evaluation of the project areas (Step 3b) in each region. From its assessment of the number of pond farmers in each sub-project area, the terrain, physical condition of canals and ponds, yields of shrimp currently attained, and the proximity of processing plants and ice-making plants, the team was able to rank sub-project areas according to collective criteria.

Concurrent with this general evaluation, the team estimated the potential adoption rate of the project by farmers, and considered constraints which might prevent them joining the project and the incentives and assistance required to help them overcome these difficulties (Step 4). To assist the team to identify all people who might be involved in the project (Step 4a), the marketing expert prepared a schematic figure of the distribution channels for fish and shrimp quantifying, whenever possible, the volumes of fish passing through the channels and the price mark-ups at each stage. Before this analysis some of the team expressed concern that the provision of support to farmers to grow shrimp for export would result in a reduction in the supply of milkfish, produced in the same ponds, to local markets. The analysis, however, indicated that the price of milkfish was sufficiently high to keep farmers supplying this sector of the market.

All these evaluations (Steps 3a and 4a) and studies (4b) were carried out by the team concurrently but, nevertheless, the team was constrained by lack of time. No use was made of rapid rural appraisal techniques to establish agents' needs and anxieties about the project. In the limited number of discussions which took place with local farmers, leaders of cooperatives, and with local officials of the DOF, the team nevertheless became confident that there would be a high adoption rate. Other than physical constraints, such as the availability of water and depth of ponds, (identified in Step 3b), critical constraints were the low and erratic supply and high cost of shrimp seed, and poor technical knowledge of shrimp farming among pond farmers.

Because of the shortage of time, the team gave little attention to the need of farmers for credit. The team's working hypothesis that farmers would need loans to improve their ponds and purchase inputs was confirmed in discussions with local bank officials in each of the project sub-areas.

The team noted also that marketing arrangements for shrimp might be improved. There was a concentration of buying power in the local processing plants which sometimes held a monopoly position. However, prices obtained by farmers appeared to be reasonable, and this issue was not considered to be a significant constraint. Overcoming it could be left to later.

The team identified a number of possible courses of action (Steps 3c and 4c) depending on the number and location of hatcheries and support facilities, such as feed mills. It then ranked them in order of priority for each sub-project area, and then again ranked the project areas. In addition to the civil engineering work required in each sub-project area, and the hatchery and feed mills, each of the sub-projects included extension services and credit.

The team now considered what could happen in the project area if the project were not implemented (Step 5). In the absence of government intervention, the team concluded that there would not be a significant increase in yields of shrimp and, in the absence of supplies of shrimp from ponds many of the shrimp processing plants would have to close. Furthermore, the value of shrimp which was produced would probably be reduced by poor quality post-harvest control given the difficulty of introducing and operating a collection system over a wide area.

The team made preliminary estimates of the costs and benefits related to each of the sub-projects (Step 6b) and, after analysis (6c), recommended a possible project for financing (6d). As part of the interim report, the team recommended that a national institution should carry out detailed topographic, soil, and water quality surveys (for Step 7) in the selected sub-project areas prior to the commencement of Phase III.

This part of the formulation process had taken three months, of which one month consisted of field work and two months of report writing. (This was the time allowed for project formulation in the budget.)

Project 3 - A Shellfish Farming Enhancement Project in Southeast Asia

At the end of preparation for project formulation, a six-member team assembled in Singapore to undertake the one-year formulation work designed to increase mollusc production and improve its sanitary control in the region.

The team spent the first four months touring the countries concerned, reviewing sectoral backgrounds, likely project areas, and holding meetings (Steps 3a, 3b, and 3c). Possible courses of action, when drafted, occupied 15 pages. The team identified seven species of molluscs for attention, eight first priority geographical areas for production, together with components in research, training, production, extension, sanitary control, depuration, processing and marketing, including a world-wide study of potential export markets for various types of processed products.

Analysis and diagnosis (Step 4) took another eight months. Instruments, targets, and regulator agents (4a), although the same in general in all countries concerned, were different in detail from one country to another. Countries were grouped, therefore, into those with common features. This was complex and took more time than planned. Less time was available, therefore, for analysis (4b), which consequently was not as detailed as desirable. It was clear, however, that the key limiting factor of the project was not biotechnical capability to produce more (in terms of areas, productivity, techniques), or financial and human capabilities (in terms of access to money and know-how), but acceptability of products by consumers both within the region and outside. Distances between areas of production and markets meant that sanitary and quality control were vital for development of consumer confidence in the products. Consequently the courses of action identified by the team (4c) concentrated on ways to achieve this goal.

Assessing the future "without the project" (Step 5) had to be condensed into two months work and was relatively superficial in character. The conclusion was that supply, in the without-project situation, could not be expected to increase by more than an average of 2% over demand brought about by population increase (with variations from one country to the other), and that exports would remain small.

Because of differences between countries, the outline specification for a possible project (Step 6) was an aggregation of six semi-independent national sub-projects, with export components in Malaysia and Thailand only, together with a regional coordination project. For the coordination project, two options were identified, one with two variants and the other with three, depending on the possible combinations of national sub-projects and extent of EEC involvement in both the coordination project and national sub-projects.

The procedure had taken all of the 15 months estimated.

Discussion Topics

Points for discussion concerning the above illustrations might include the following:

- the factors which contributed to any change in project concept between Phase I (Preparation for Project Formulation) and the conclusion of Phase II (Reconnaissance and Preliminary Project Design);

- the extent to which formulation teams should be flexible towards the original project concepts;

- the extent to which the procedures for Reconnaissance and Preliminary Project Design adopted by the teams varied between the projects illustrated;

- the advantages/disadvantages of having a fixed timeframe within which to work having regard to the projects illustrated;

- causes of concern at this point with regard to each of the projects illustrated;

- any weaknesses in the procedures which may have occurred.

3. PROJECT PREPARATION

Project preparation is the second stage of formulation. It has three Phases (III-V), project design, analysis of expected results, and project documentation and submission.

PHASE III: Project Design

Purpose and Outputs

Project design starts after the project sponsor has reviewed the formulation team's interim report and agreed that work on the project should proceed to the design stage.

The purpose of project design is to plan the project to an appropriate level of detail in terms of the technical specifications, organization structure, scheduling, management arrangements, resource requirements, and cost estimates, in order first, to provide the project sponsor with the information required to make an investment decision, second, to plan the project for implementation.

As noted at the end of Phase II, in private sector projects the project sponsor, by this stage, might have received a prefeasibility study report on which he could invest in the project immediately, and all that would remain would be the planning for implementation, provided local approvals had been obtained. In larger projects the project sponsor may have yet to decide on certain design options and then secure the appropriate approvals, and possibly seek further financing. Public sector projects invariably have to develop the preferred development option much further in the design stage before financial assessments are made.

There are five main outputs of this phase:

- Project objectives and targets are specified in detail.
- Project components are fully designed.
- The overall project phasing and scheduling of individual components are determined.
- All required organization structures and management arrangements are fully identified.
- Capital and recurrent costs are estimated and a financing plan is worked out.

In project identification the team was engaged in the collection and analysis of information to identify opportunities and, through further analysis, to arrive at the proposed project. This phase, however, is iterative in character, the team checking in more detail the practicality and the probable effectiveness of the preliminary design proposals. In Phase II the team had to ask itself, "Do we understand what needs to be done?" In Phase III the question is, "Are we sure how it will work?"

Typically work is broken down into five steps (Steps 7-11), the contents of which are summarized in Figure 7 and described below.

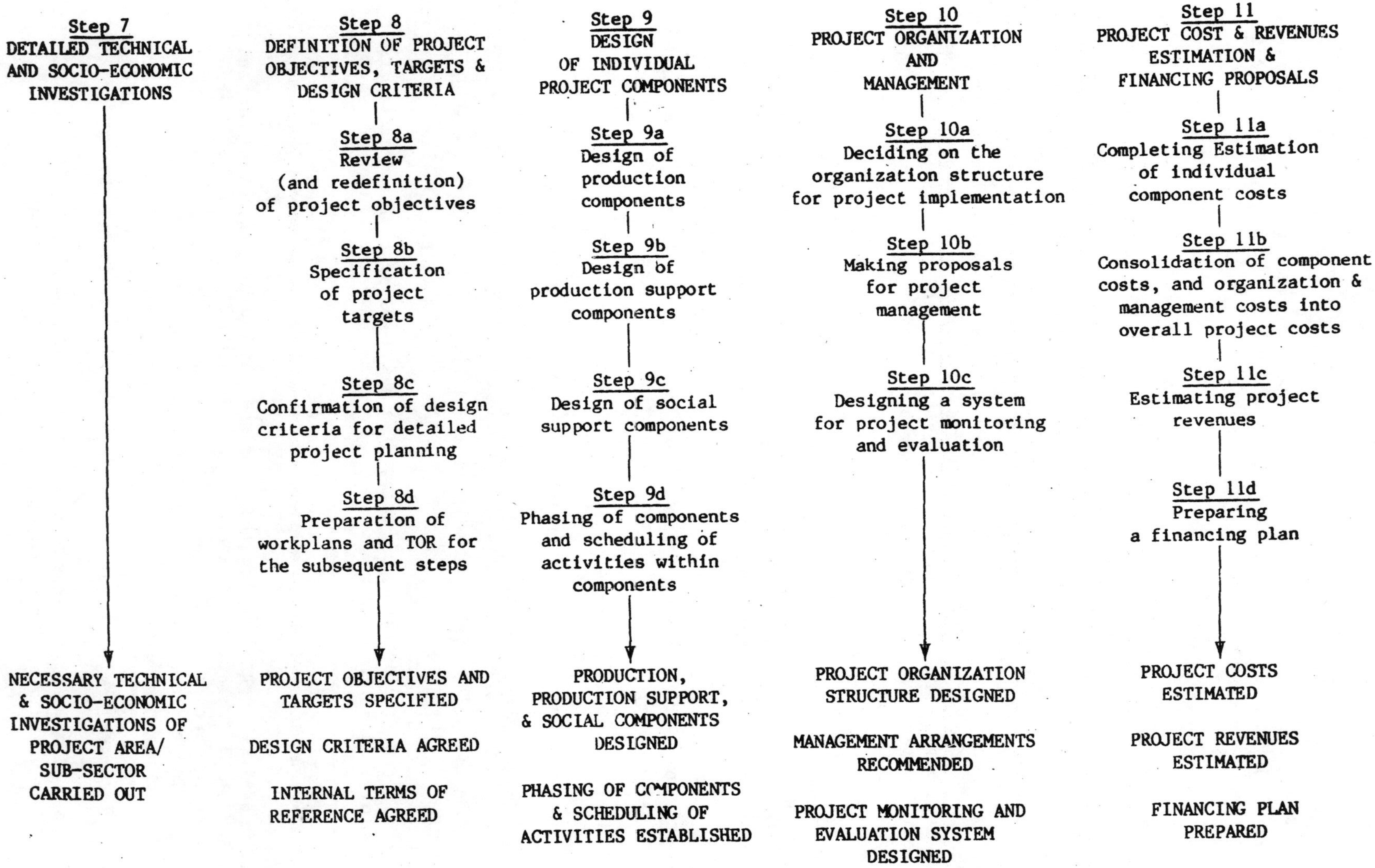

Figure 7 Phase III. Project design

Step 7 - Detailed Technical and Socio-economic Investigations

This Step is mostly relevant to public sector projects where more specific issues have to be answered in response to the preferred development option.

After a project has been designed on a preliminary basis, it is often necessary to obtain more detailed information on certain technical and socio-economic aspects from specialized field surveys. Normally these surveys will have been recommended by the formulation team in its interim report.

Although these surveys would, within the framework of project identification and preparation, be seen as typical activities in preliminary project design and should take place within the identification stage, frequently the reason for their deferment in public sector projects is that they are often costly and, until identified in the interim report, the costs have not been authorized. These surveys are usually referred to as "pre-design studies". The work is separate from the project design phase and is usually completed before any further step is taken. In most cases surveys are performed by teams independent of the project formulation team, working from TOR established by the team.

Typical surveys in aquaculture projects include the capabilities or limitations of government departments, marketing boards, supply companies, and the types of agents who will be concerned. Of increasing importance are socio-economic and environmental surveys. The former is necessary in many production projects to ensure that the right instrument agents have been identified, that social constraints and opportunities have been clearly delineated, and to obtain sufficient information about the social structure so as to secure the involvement of beneficiaries in project implementation and after. A socio-economic survey may also be required, for example, to establish the costs and earnings and the farm systems employed by small farmers in order to determine the appropriate fish farming systems and their likely attractiveness to small farmers in terms of the net earnings which will be generated, level of risk, etc. Environmental surveys are now almost indispensable and may include, for example, disposal of farm effluents, destruction of ecosystems, or the genetic diversity of natural populations. Also included may be surveys of legislative and other requirements, as in some countries it may take a long time (and involve certain costs) to obtain the necessary permits and licences for operations.

It may also be necessary at this stage to set up market studies from which the results may not be available until later in the design or analysis phases. Alternatively, market studies may be undertaken by a marketing expert as a member of a preparation team during the project preparation mission.

At this stage also, where projects are likely to be dependent on inputs, such as feeds or seed, it is often desirable to establish the conditions under which such inputs will be available.

In production-oriented projects where the selection of sites has to be narrowed down, the surveys may include further site quantification, building on the work carried out in Step 6b.

Surveys may be so extensive or important that it may be decided to postpone the subsequent steps in project planning until the results are available.

When all activities are completed the project formulation team should have available all the information required to carry out project design.

In private sector projects almost all of this work will have been carried out in Steps 4 and 6. Exceptions may be larger projects which have had a broad mandate through reconnaissance and preliminary design. An example is an international company interested in the potential for converting several of its unprofitable sugarcane plantations into aquaculture but not knowing what was feasible. At the end of Phase II the project team might have recommended two plantations as having the greatest potential. The company might then approve the continuation of the work through the design phase in respect of one site.

Differences in Public/Private Sector Procedures (Step 7)

The single task of Step 7 is relevant to public sector projects (Models B, C, and D), and may be pertinent to the broader private sector projects (Model A) which have options to analyse in more detail.

Step 8 - Definition of Project Objectives, Targets, and Design Criteria

This step may occupy only a short time in the formulation team's schedule but it is an important aspect of the formulation process. It marks the end of the main period of data gathering and the beginning of actual project planning. As with Step 7, the tasks of Step 8 are mainly pertinent to public sector projects; in the private sector, the tasks are relevant only to those projects which started with broad objectives and had few specificities with regard to location or species.

Step 8a - Review and redefinition of project objectives

The interval between the submission of the interim report and the start of work on project preparation provides an opportunity for the project sponsor and/or the national planning authorities (and the formulation team) to consider the proposed objectives of the project and its preliminary design.

In this period any of the interested parties may propose modifications to the objectives and the preliminary design. When this occurs, the suggested modifications should be discussed and agreed by the project steering committee, or the relevant official body, prior to the team redefining the project objectives.

This is the last opportunity to review and revise objectives.

The redefinition of objectives may consist of, for example:

- a fully detailed statement of the project's intended direct effects (such as, increased yield per hectare, significant increase in a number of small-scale fishfarmers); and indirect impacts (such as, incomes for rural households grouped according to their characteristics);

- indications of the approximate magnitude and timing of the project's primary outputs (such as, tons of fish produced by a given year).

This task is particularly important in projects where objectives are quantified into targets. In setting them, the team should think through the entire project process. For example, in meeting a target for trained extension agents the team should take into account the numbers of training instructors, the volume of training materials, the logistics of space, and rates of attrition, etc. In production projects, to meet an objective of X tons of fish or shellfish in a four year period, the team may have to consider the time required to build up the amount of broodstock needed to reliably provide seed to meet production forecasts. When setting the objectives, therefore, the team should be confident that it has the necessary background information. This information should have been obtained during project identification (Steps 3 and 4) and in Step 7 of this Phase. If it has been missed, the team should take time here to collect it.

Step 8b - Specification of project parameters

Decisions must now be taken to establish the four principal parameters of the project. The order and the content of work required to reach the point where decisions can be made for each varies with the type and focus of each particular project. As noted in the chapter on project identification, for projects which are based on a specific idea at the outset (Models A & B), much of the work in this task will have been completed or partially assembled. For others, particularly the general public sector projects (Models C & D), this will not be the case.

The four parameters are:

(i) Demarcation of the project area/zones. For production projects where the precise site has not been identified, the team should undertake a complete quantification of all remaining design options and the bioprogramming components described in Step 6b. For example, in a project with the objective to increase production of molluscs, a number of coastal areas may have been identified in the interim report but final selection of sites within those areas might have depended on (a) water quality and circulation surveys, etc., carried out in Step 7, and (b) bioprogramming to determine potential yields of each site, and costs of supporting services (construction areas, depuration plants, processing plants, etc.).

For other types of projects, such as establishing a regional training centre, similar considerations would apply. In this case the team would follow the bioprogramming formula (substituting numbers of trainees and their needs for biological criteria), and develop the infrastructure requirements and costs.

(ii) Selection of project agents. For the most part agents will have been identified and selected in Step 4c, when the project location has been relatively well defined. If, however, the final choice of the site has been dependent on surveys and studies (Step 7), the team must now select the instrument and regulator agents for the project.

One of the problems which may be encountered in production projects is that, in practice, it may only be possible to estimate tentatively the number of producer agents who will participate in the project (the adoption rate). This problem emphasizes the importance of the formulation team having a thorough understanding, by this time, of the local social and economic structures in the project area (Step 4b), and the "identity profile" of individuals most likely to be attracted by the type of farming activity proposed. The problem is less likely to occur in non-production projects, such as research and training, where selection of instrument agents should not present any difficulty.

(iii) Confirmation of selected technologies and management practices. This is only applied by the formulation team in public sector projects and concerns the project at farm level. It mostly confirms that the proposed technology meets the needs of the producer agents, in particular its reliability under most conditions, its micro-feasibility and compatibility with farmers' resource base and management constraints, its incentives and attractiveness to users, and its wider implications and affect on agents other than the farmers.

In projects which are specific with regard to location and species, etc., the technology will have been decided in project formulation, and no further work is required. For general projects, options for implementing production may still be dependent on final site selection and specific agents. For example, one site for increased mollusc production might lend itself to raft culture as the location is deep; at others production on the substrate may be more practical because the offshore location is shallow. Even so, these physical conditions may be offset by considerations relating the selected technology to the needs of the target agents. Where the agents are already familiar with raft culture the team may decide to select this technology in preference to utilization and management of mollusc beds on the substrate, even in a shallow area.

This type of judgement is not always easy, as aquaculture as a sector does not have the long traditions which permit ready comparisons of one type of farming technology with another. Relatively easy calculations can be made to select which system is the most feasible, for example, selecting between extensive, semi-intensive, and intensive systems, as the availability of resources is often the deciding factor rather than pure economics. However, choosing the right practice, such as selection between farming with floating rafts and management of seeded beds, or between earthen ponds, cages, and net-pens, or even choosing the combination of species, are much more difficult problems. Good decisions depend upon a thorough knowledge of the technologies available and of the people involved.

(iv) Phasing and timing of project activities. This task is relevant to all projects. It is not concerned with the scheduling of the project, which is undertaken later (Step 9d) but with understanding and properly estimating the degree of difficulty, both psychological and physical, which changes proposed by any project may represent to the instrument agents concerned. At the end of project identification the broad thrust of most projects will have been defined. Here the team has to confirm that phasing and timing of the project, and its support components, conform with the direction of change envisaged.

The team might be assisted here by reference to the following set of standard guidelines which were prepared originally for agriculture development projects and are adapted here to aquaculture projects.

Characteristics of Options for Change

- Projects which only require adaptation of an aquaculture technology can often achieve their objectives in the short term. Adaptation describes adjustments of any existing production system without significant modifications. As aquatic farming is very site specific, adaptation of production technology is required by almost every production-oriented project. Production practices based on extensive systems are readily adaptable as they require few inputs; semi-intensive systems invariably require adaptations for supplementary feed formulation and seed production, and it may take longer to achieve a project's objectives; finally, intensive systems, because they are more technically advanced and sophisticated, may require prolonged and difficult adaptation.

- Projects which require expansion of an existing technology over a greater area can achieve their objectives usually in the short or medium term. In many cases, where profitable farming is taking place, it will be replicated by the private sector without government intervention. In some instances, however, it will be difficult for farmers to convert their existing land or take up new land without support from the government in the form of access to credit, provision of infrastructure, and support services. Some adaptation of technology will also be required to suit the new areas, which will also impose a delay.

- Projects which require reorganization, such as significant changes in the production system and marketing of the product, can achieve their objectives in the medium to long term. If, for example, a project intends to convert farmers producing fish for local consumption into farmers producing shrimp for international markets, or a project encouraging displaced lake fishermen to engage in cage farming of fish, it is clear that the socio-economic systems in which they operate will have to undergo a radical change.

- Projects which introduce the technology of aquaculture can achieve their objectives only in the very long-term. This is because inevitably there is a slow adoption rate, and for an economic activity to become viable within an economy it has to become of sufficient importance to attract an appropriate level of government support and services from the private sector. For example, the salmon industry in Chile, which first began in the early 1970s, cannot be said to have reached a level of self-sustainability until the late 1980s. During this period the industry received continuous financial and technical assistance from Japan; Fundaçion Chile provided continuous support for research and development, demonstration farms, and marketing; the Government provided fiscal incentives and the appropriate legislative framework; and major commercial interests in the USA and Norway made long-term investment.

Step 8c - Confirmation of the design criteria for detailed project planning

This task, as with the two previous ones in this step, is mostly applicable to public sector projects and to a few broadly defined private sector projects.

In Step 1c the Project Sponsor would have established certain criteria for the proposed project. The team leader should now confirm that these criteria continue to stand as drafted, or whether they have been modified following the submission of the interim report. Once all the appropriate criteria have been established for the last time, the team leader should ensure that each team member is familiar with them so that their work in the design of individual project components (Step 9) is consistent.

Project criteria can be organized conveniently into the following categories to assist this process:

- Financial criteria (costs, cost recovery, cash incentives, etc.).

- Economic criteria (minimization of resource use, maximization of outputs for which there is a high demand, etc.).

- Social criteria (improved access of agents to resources and services, enhancement of self-organization of the sector, etc.).

- Environmental criteria (protection of essential biological ecosystems, such as mangrove swamps, effect of effluents on water quality, etc.).

- Implementing criteria (complexity and difficulty of tasks versus capabilities, attractiveness, sustainability, etc.).

Step 8d - Preparation of workplans and TOR for the subsequent steps

After the (re)defined objectives and activities of the project have been discussed, and the major criteria have been agreed, the team leader should prepare TOR for each person involved in Steps 9, 10 and 11, and schedule the work. These TOR should:

(a) define individual responsibilities in terms of:

- specific outputs,
- specific reporting format,
- deadline for submission of the first draft, and

(b) identify all interfaces between:

- team members,
- relevant organizations, groups or individuals outside the team.

Even at this point the team leader should be looking ahead to the preparation of the <u>final project report</u>, and will have determined the main parts of the report, working papers, and technical appendixes. This may best be done by preparing a complete <u>Table of Contents</u>. It is often good practice for the team leader to provide the team members at this time with examples of the type of work which is expected, such as the detail required and its style of presentation.

Differences in Public/Private Sector Procedures (Step 8)

The first three tasks of Step 8 are entirely applicable to public sector projects (Models B, C, & D), and only those private sector projects which began with broad or general objectives. The fourth task (8d), is applicable to all projects going into design (Step 9).

Step 9 - Design of Individual Project Components

As a prelude to Step 9, it is useful for the formulation team to separate components into categories before individual component design. These categories are:

- Production components, such as facilities (farms, hatcheries, processing and feed manufacturing plants), which directly create added value;

- Production support components, such as facilities and structures (research centres, training schools), and services (research, credit, training and extension, veterinary advice), which enable the production components to add value;

- Social support components, such as infrastructure (housing, potable water, meeting facilities), and services (education, health services, community groups, savings and loans schemes), which are not essential for economic success in the same way as other project components but which contribute to improving the quality of life in the project area/sub-sector. These exist only to a small extent in aquaculture development.

- Technical assistance, which is a support service of limited duration but is treated conventionally as a separate component in project formulation (Step 10b).

The project design should provide sufficient information about each activity so that it serves as a comprehensive reference document, and predicts, as accurately as possible, costs and their timing. However, as with Step 6, there is an important difference between the level of design of each component for projects in the public sector and the private sector.

In public sector projects of a general nature, for example a project to support small-scale pond farming, the design of most support components, social support components, and technical assistance components is not carried out too rigidly because, as has been noted, there are uncertainties attached to some activities and some flexibility during implementation is desirable. Progress in project implementation, for example, may be dependent on the adoption rate of technology by both existing and new small-scale farmers. In such cases it will be possible for the team to provide only approximations of progress and the resources required over time.

Even with public sector projects which involve the design of production components and some support components (facilities and/or structures), the level of detail is often dependent on the project sponsor. Donor organizations, for example, usually require only "preliminary designs" (diagrams or sketches of lay-outs), to be prepared together with general lists of equipment/machinery, grouped in main categories.

In private sector projects, however, this Step is almost entirely concerned with "final design" of production components and possibly some support components, and estimation of the costs with a high degree of accuracy. Therefore it is necessary in order to prepare final design drawings and bills of quantities. These may not, in fact, be prepared by the formulation team, if it has no professional architects and engineers and the project sponsor may contract this task to a professional company.

Whatever the project and its components, the level of detail of design required should be specified in the TOR (Step 8d).

Step 9a - Design of production components

Production requires facilities and/or structures. Therefore, the design of production components must involve the input of professional designers to make certain the components function properly, meet all standards for construction and are safe to operate.

The design of production components is usually required in three types of projects. In brief, these are as follows:

(a) almost all private sector projects (Model A);

(b) public sector projects which include a publicly owned facility, such as a state hatchery, demonstration farm, production farm, or a research and development centre (Model B);

(c) public sector projects which will require investment by private farmers in facilities (Model C).

In projects described in (a) and (b), above where facilities are to be provided, the quantification and preliminary analysis would have been completed in Step 6. The basis for the work of the architects and engineers in this Step, therefore, are the prefeasibility study report and the bio-programming document (if it was a separate report), which were produced in Steps 6b and 6c.

The work may divided into two parts, namely:

(a) preliminary design, and
(b) final design.

This is because the design and costs can be narrowed down further depending on the answer to some structural alternatives. For example, whether the hatchery should be built of brick or wood; whether the raceways should be constructed above ground or below; whether the research and training centre should have classrooms and laboratories in one building or two, etc.

In preliminary design major emphasis is upon the civil, mechanical, and architectural design of each facility. The design of all mechanical processes (such as the water systems) will be completed at this time to ensure that they are integrated into the structures correctly. Architectural concepts are developed, and structural systems identified. From information collected during bioprogramming, criteria are developed by the engineers for such factors as soil compaction for berms and dikes, and water treatment. A sufficient number of preliminary design drawings are completed to communicate the design concepts, and enable the costs to be recalculated.

In final design the "blueprints", or detailed architectural and engineering drawings are produced for all physical components, and may be accompanied by a final design report. This enables the costs to be recalculated yet again with greater accuracy and approximate actual construction costs. Virtually all design problems will have been solved before final design, and changes during this stage usually cause cost over-runs.

In projects described by (c) above, namely public sector projects which require private investment, this task usually involves only preliminary design. This enables common standards to be set for all facilities or structures to be built through private investment if credit is to supplied by the project. For example, in a public sector project to increase the production of freshwater prawns, the project might involve a central government hatchery, which is carried through final design, but only preliminary design and specifications for grow-out ponds, such as designs for 0.5 and 1 ha ponds with standard water inlets and outlets, and berms of minimum width with standard compaction and slope.

Finally, in the private sector final designs and specifications are in such a form that preliminary bids can be solicited from contractors and suppliers to support project cost estimates. In the public sector designs and specifications remain generally less detailed, and cost estimates are made by reference to price lists and enquiries.

Step 9b - Design of production support components

Excluding support components which involve facilities and/or structures, the design of which has been described above, the majority of production support components relate to services such as extension services, training course, study tours, credit, veterinary advice, etc. These are usually less common to private sector projects than to large multi-components projects in the public sector, particularly those supported by international assistance. Most donor agencies have specific guidelines and policy booklets to assist the formulation teams in this Step.

The formulation team should keep in mind a number of key points when dealing with production support components. Firstly, the design of many support components must be made in close relationship with that of the corresponding production component; for example, any proposals for institution building should be made in relation to the projected size of the sub-sector. In some cases production support components may account for a major part of the project cost, and it is important to find the proper balance between minimizing their costs and ensuring that they fully meet the needs of the project. With regard to extension services, training, and technical assistance, for example, it is important (but difficult) to estimate the level of input of each component which will be required to produce or increase a unit of output in the respective production component.

Step 9c - Design of social support components

Many regionally-based rural agriculture development projects are frequently used as vehicles for expanding the scope and upgrading the quality of social services provided to rural communities as there may a direct link between social support and production components. For example, some irrigation projects may include the addition of a system to provide clean drinking water to village communities, or an extension centre may also be used a a centre for information on hygiene and public health.

There has been little opportunity to date to demonstrate where and how aquaculture projects in the public sector might include social support components, other than sharing facilities or possibly supplying potable water to a small community.

So far, therefore, there has been little risk of projects becoming difficult to manage and over-complicated through their inclusion. Project planners, however, should be aware of this issue and take care that, when opportunities occur, there is a reasonable balance between satisfying the basic needs of the target agents and meeting the objectives of the project.

Step 9d - Phasing of components and scheduling of activities within components

Toward the end of Step 9 it is necessary in public sector projects to bring together the completed and semi-completed designs of the various components to consider their phasing and scheduling. Phasing and scheduling may also be considered for some large projects in the private sector, in order to plan for financial returns on investment without expending all the capital or waiting for the complete facility to be built.

For relatively simple public sector projects with one or two components, phasing should not present any particular problem. However, in the case of large multi-component projects, decisions on phasing can be critical.

A logical sequence of activity in a complex project might be as follows:

- A start-up phase of detailed field investigation, pilot-scale testing of new infrastructure, equipment and technological changes, and organization of participants. This may last from six months to two years.

- A first operational phase, with introduction of key technology support packages, adequate production support components and institutional strengthening, with minimum supporting infrastructure. This may last from one to three years.

- A second operational phase, with complementary technology support packages (building on the first or diversifying), and further investment in infrastructure. This may also last from one to three years.

- A completion phase, with emphasis on community and social services.

In most cases, project implementation will lag behind the schedule set out in the project formulation documents. The frequency with which this occurs suggests that the fault invariably lies more with an <u>unrealistic plan</u> than with inefficient implementation. For example, in many countries the public authorities have elaborate procurement procedures which project management must observe. Conversely, the team has to guard against drawing up an implementation schedule which takes full cognisance of the procedures of the implementation agency but where no account is taken of those of the financing agency. The scheduling of specialized personnel inputs for various activities requires particular appreciation of the logistics of recruiting, placing, and supporting this type of manpower; slow recruitment is a very frequent cause for delay. Moreover, there are unavoidable seasonal constraints which have to be observed in both administrative procedures (such as fiscal years and annual budgetary commitments) as well as in the practicalities of working (such as the difficulty of construction in the rainy season). It is not difficult, therefore, for a few weeks of delay in implementation to be incurred which can often end in the loss of almost a full year if, for example, the start of a breeding season is missed.

It is essential, therefore, that the team avoids excessive optimism about the time required to start up a project and implement it. The various forms of network analysis, for example, PERT (Programme Evaluation and Review Technique), are of value in these situations. Formulation teams often do not possess the skills to carry out PERT, while many aquaculture projects will be relatively small and will not require such a sophisticated technique. They do require, however, a more systematic approach to the problem than merely an intuitive process, which is frequently what takes place.

The team should, first, take into account the performance of the implementing agency with other projects, where possible discussing this with the donor agencies concerned to identify where delays might be incurred; second, it must analyse the activities required to implement the components, and to sequence them in a bar chart. Most projects contain a number of activities, such as building a hatchery, training extension workers, setting up a credit scheme, etc. When activity schedules have been completed for each one they should be integrated into a bar chart for the project as a whole. An important part of this process is to identify and build into the bar charts the prerequisites of any particular activity, for example, land purchase for the hatchery.

Differences in Public/Private Sector Procedures (Step 9)

The four tasks of Step 9 concern private sector and public sector projects to different degrees. The design of production components (9a) is common to projects described in Models A, B, and C. If the components involve facilities and structures the design requires completion of professionally prepared preliminary design documents, and in most cases final design blueprints. Similarly if Models A, B, and C include support components (9b) which involve facilities and structures then the same also applies. Public sector projects (Models B, C and D) are more likely to have several other support components, such as extension and training. Social components (9c) are almost entirely applicable to public sector projects (Models B and C), but are not readily identified for aquaculture. Private and public sector projects described by all four Models require phasing and scheduling of activities (9d).

Step 10 - Project Organization and Management

Organization and management are related aspects of administration. Organization is essentially about structure; management is essentially about operations.

The organization and management of a project should be designed with the long-term objective of successful institution building, so that the project will continue to operate satisfactorily once all temporary supports (such as technical assistance and external funding in public sector projects, and fiscal assistance and parent company support in the private sector) are withdrawn. In other words, good organization and management help to ensure that the project is self-sustainable.

Organization is concerned with:

- Allocation of responsibility for specific tasks among the various participating bodies.

- Establishment of hierarchical structures for decision-making (including definition of individual limits of authority).

- Arrangements for the proper exchange of information on project activities.

Management is concerned with the planning, implementation, and control of all activities which come within the defined jurisdiction of each body participating in the project organization.

As organization and management are often both politically and technically sensitive the team should adopt the following general approach to its task:

- Familiarize itself with existing organization and management aspects of development administration in the area/sub-sector concerned.

- Draw up its proposals for project organization, management, and evaluation in provisional form.

- Discuss these provisional recommendations with the principal agents who will be affected to reach a consensus wherever possible. This does not necessarily mean that the team should adopt all comments, but that all key agents are given the opportunity to express their points of view.

There are major differences in the organization and management of projects between the public and private sectors. In public sector projects the organization and management has to conform and fit in with existing organs of public administration. In the private sector project sponsors can make whatever arrangements they consider to be most effective, having regard to the size and complexity of the project, and their own financial and technical capabilities. The forms of organization available to the private sector were briefly described in Part I, Chapter 2.

The description of the team's work in Step 10, therefore, is confined to public sector projects.

Step 10a - Organizational structure for project implementation

The work of the team in this task is to decide which entity or entities should be responsible for project implementation and operation, and how they are expected to perform these functions.

General organization. For sub-sector based projects the team will usually find that the organization most logically suited to implement and manage the project is already concerned with development in the sub-sector, such as an agricultural development bank or the Fisheries Department.

For area-based projects, unless there is a regional agency which will assume sole responsibility for project implementation, the choice of an organization for project implementation may be more difficult. This is especially so where large projects have to be implemented and more than one existing government institution and perhaps parastatal bodies try to play a role, or where their participation is essential. For example, institutions interested in a typical aquaculture project may include the Department of Land and Water, the Department of Fisheries, Ministry of Agriculture, and the Agricultural Development Bank. In this situation, the team will have to find a way to coordinate (and resolve any differences) of these leading institutions at the higher policy making level, where major issues, such as land and water allocation, have to be resolved. It may be appropriate to organize this coordination through a committee of ministerial/departmental representatives with delegated authority to resolve policy issues, but which meet infrequently.

This situation has to be distinguished from the coordination required at the day-to-day level among the key agents, such as farmers, cooperatives, local government, district officer of the central government, and all others who are involved in the project (see Step 4b). Here, the team may find that the best organization is also a committee which operates at the district level (or its equivalent), but which meets very frequently.

However, when approaching the issue of organization through committees, the team should ensure that there are clear lines of responsibility, and leadership and management of the project is not diffused. This means that the team has to address itself to the issue of possible competing and conflicting aims when drawing up the TOR for these committees, and obtaining the agreement of the relevant institutions and key agents.

It is partly because of organization and coordination problems perceived within existing bureaucracies that formulation teams sometimes advocate an alternative form of organization, namely a parastatal body or project authority financed and managed independently. This approach parallels that of the private sector when the parent company sets up a new company or subsidiary for the project. However, for aquaculture projects, which are usually quite manageable even when large, this approach is not recommended except when the existing structures are totally inadequate. Building a new organizational structure is often extremely difficult and time consuming itself, and there is always a potential risk of conflict with existing structures if there are differing arrangements, such as staff benefits and preferential procedures.

Organization of project beneficiaries. The issue of securing beneficiary participation in the design of those projects which reach a large number of beneficiaries (in other words, a high adoption rate by farmers) has been referred to above (Step 4b). There were two purposes for the participation of farmers at that time in the planning of the project: (a) to determine their views and priorities, and the feasibility of different project options; and (b) to find means of securing their genuine involvement in project implementation and after. The success of a project where the adoption rate is critical will be dependent, to a greater or lesser extent, upon the organization which the team sets up, based on its discussions with farmers in Step 4b, to encourage farmers to become involved. In agriculture, these typically are users' associations (for equitably sharing resources, such as water), marketing associations, and simply associations of farmers. These associations have valuable functions, namely, to provide focal points for the project and thereafter a channel for communication between the responsible government department and farmers, and to enhance the ability of farmers to solve their own problems by using their own resources.

However, when fish and shellfish farming are new activities, the creation of these associations is not possible until there is a large number of commercial farmers. By this time the project will often be over. In these cases the organization of the fish farmers might be made through existing associations in the agriculture sector. When considering this role of existing associations, the team should determine their ability to solve their own problems with their own resources, and to identify further support they may require to service the fish farmers effectively. Then the team will be able to adjust their proposals or evaluate their needs properly to integrate with any new organization.

Where an aquaculture activity is a component of a rural development project in which the target group is made up of the more economically deprived and socially weakest members of the community, it will not be possible for the project to take advantage of existing associations. An important part of the project in this stuation, therefore, is to identify a focus or issue upon which an association can be built, such as the equitable sharing of water or the marketing of the farmed products. However, as pointed out in the introduction to Part II, there are few such projects at the present time.

Where fish or shellfish farming is established, and there is already an existing number of farmers but without any association, the focus of any new association will perhaps be the resolution of common problems, or a specific purpose such as marketing. In this case establishing such associations is likely to be easier.

Ideally, farmers' associations should be financially self-sufficient (through members' subscriptions or levies on transactions), but in many cases this situation may take some time to bring about.

Relevance of project financing arrangements to organization proposals. This task only applies to larger aquaculture projects where certain activities require financing prior to them becoming financially self-sustaining, or where the government accepts the burden of financing after completion of the project. For example, where projects are dependent on the installation of central pumping systems and distribution canals, the pumping costs will usually be carried within the project budget until they are assumed by another entity on project completion. Another example might be a project hatchery, constructed and operated with project funds to provide seed as part of the support to potential farmers, which is to be sold to the private sector at the end of the project. In such cases the team should be clear about the flow of project funds through the administrative system of the project.

One important point for the team to bear in mind is that individual officers and committees, who exercise a financial control function and thus have responsibility for authorizing expenditures for the type of activities noted above, or for activities at other levels, often in practice take upon themselves a decision-making rôle about technical matters, regardless of their intended function and technical competence in those areas where decisions are required.

Step 10b - Proposals for project management, including training and technical assistance

Relationships between management skills and project objectives. Poor standards of management are the largest single source of institution problems faced by projects in the agriculture sector. There is no reason to believe that aquaculture projects are significantly different in this respect.

In many instances there is little connection between the care taken in the project design phase to ensure that the management arrangements are appropriate to the project, and managerial performance when the project is being implemented. In many such cases, the effectiveness of management may be reduced by bureaucratic delays in the release of resources, by institutionalized corruption, or by nepotism and political interference in the selection of staff.

On many occasions, however, teams could have taken greater care in project design to ensure that, when job descriptions were drafted, together with the qualifications and experience required of key staff, the tasks required of management were realistic in terms of the skills and experience known to be available to the project.

The importance of this Step is underlined when it is recalled that one of the principal tasks of the project planner is to ensure that there is a "fit" between the project tasks, the skills and experience available, and the timeframe within which the tasks have to be accomplished. Therefore, the team has to make a realistic and objective assessment of both the existing level of skills and the potentials for transfer of management and technical skills, and special knowledge which will be acquired under any training or technical assistance provisions (see below) in the project. If, after making this judgement, there is doubt about the "goodness of the fit" then one of the variables must be changed. If the level of available skills to the project have been assessed and found wanting, attention then has to be switched to the project timescale or the project objectives. In most instances the timescale is fixed. The logic of the situation would indicate that the objectives must be revised so that the tasks to be performed fall within the capabilities of the project management.

The issue of management and other skills, therefore, has to be approached early in the preparation stage when the first approximation is made of the project tasks and components. Unfortunately, the techniques for assessing skills are time consuming and are concerned with only one of the factors affecting institutional performance. It is most likely, therefore, that in practice the team will have to make its judgement on the basis of assessment of institutional capability, giving as much attention as possible to performance in decision-making and administrative procedures.

Project management arrangements.

One of the first tasks of the formulation team in this Step is to draw up TOR and job descriptions for the project management staff which the team believes is required (bearing in mind the points noted above), and to specify the qualifications and experience required of key staff. This includes both national counterpart staff and any international experts. It should be noted here that the TOR of any national counterparts should not be identical to those of international experts, although this is too often the case.

The team also sets up the structure to indicate relationships between different management responsibilities, and the lines of authority. A typical example of an organizational chart is illustrated in Figure 8.

This is followed by consideration of the organization and physical location of the project management staff within the institutional structure.

In the case of small projects in the public sector the management staff is usually physically located and remains in the principal organization structure concerned, such as the Department of Fisheries, or the Agricultural Development Bank, but without the creation of any specific project management unit (PMU). In larger projects it is usual for have a PMU, particularly if external financing is included. This is usually a temporary arrangement and it is disbanded once management of project activities is absorbed into the regular programmes of the agencies concerned. The PMU is typically attached to and physically located within the appropriate line Ministry or, in the case of area projects, may be placed within an area development authority or other regional institution.

Training. Institution building and development to a large extent means training people. The formulation team may conclude that managerial and technical skills of the staff of an implementing agency requires upgrading, and the project beneficiaries also need technical training to use properly the services and technologies which the project is making available. It may also be necessary to include training of peripheral participants, such as loan officers responsible for disbursing the credit. The formulation team may therefore draw up a training programme involving recruitment of consultants, who may be attached to the PMU or to various organizations within the project structure, for periods of training implementation. Alternatively, the participants may be sent on existing training courses elsewhere, or on study tours.

Technical assistance. Technical assistance (TA) is a costly part of the overall project expenditure, especially if any consultants required are international experts. It is important therefore to ensure that consultants providing the TA are used to maximum effect. This requires close involvement of executing agencies in the recruitment of consultants and in the provision of counterpart staff of the right calibre.

It is important that the formulation team takes care not to provide excessive TA in an attempt to overcome weaknesses in local management and

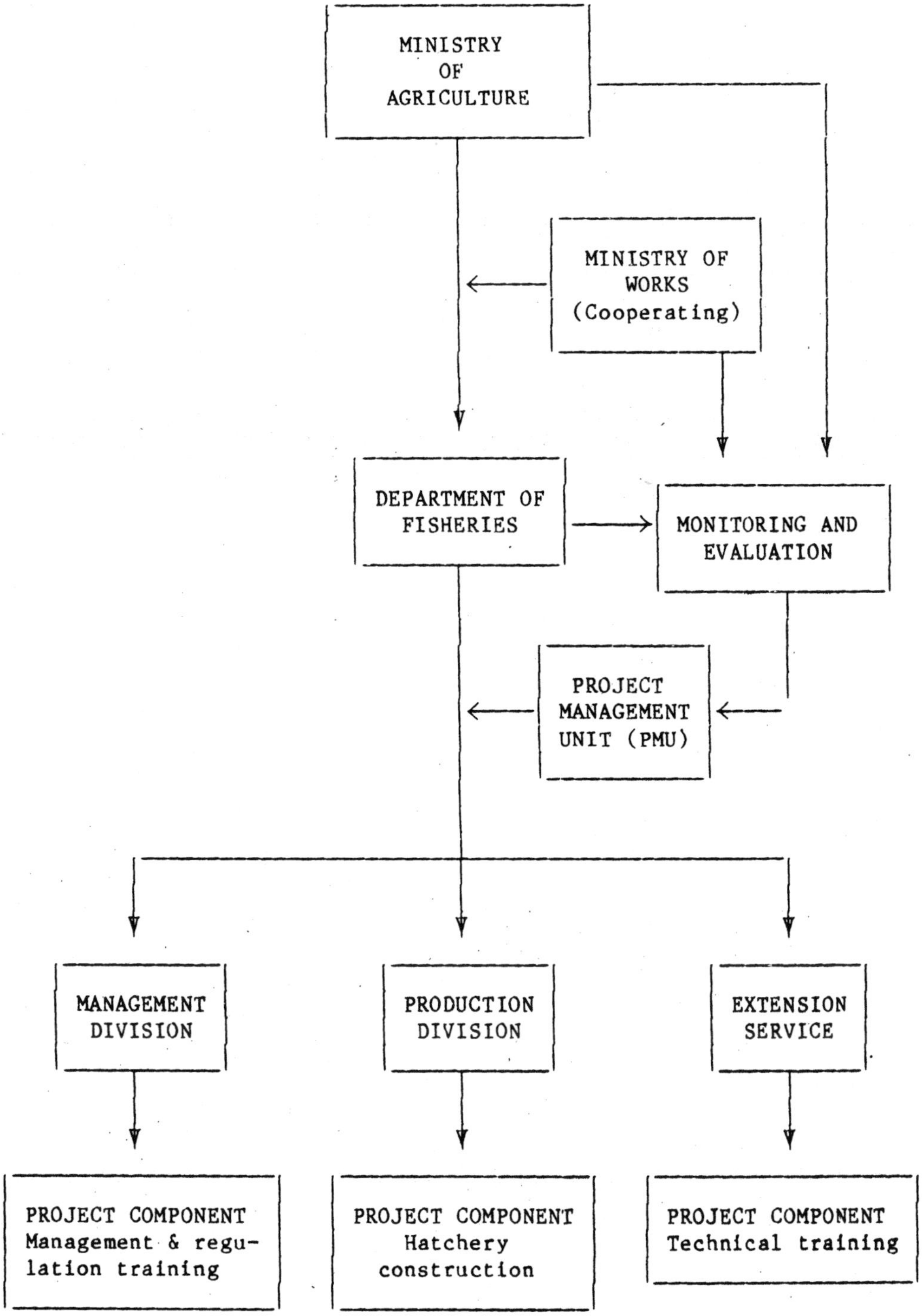

Figure 8 Project management organogram for a project to enhance inland fisheries through hatchery construction, improved management and training

technical skills. When such TA is withdrawn with the conclusion of external financing, the sustainability of the project may be jeopardized. It is preferable that the project tasks and components are so designed at the outset that national counterparts are given proper opportunities to manage the project on their own, even for short periods, and to learn from their mistakes. The team should keep in mind that it may often be preferable to use consultants in management "back-up" roles, making short visits at regular intervals rather than to appoint them to residential positions on site, even though this may make project implementation proceed more slowly and perhaps less smoothly than desired. In this context the potentially useful role of supervisory missions, by the organization which has undertaken project formulation, should be taken into account.

In the private sector, as noted in the introduction of Step 10, the project sponsor decides his own management structure and staffing needs. However, the project team can assist by preparing TOR and job descriptions for each post. In addition, when construction of large and complex facilities are part of the project, such as a hatchery, the project sponsor usually contracts the team or hires professional engineers to assist in the bid process and select the contractor (Step 11), and then provide construction supervision.

Step 10c - A system for project monitoring and evaluation

Monitoring and evaluation (M & E) are closely related activities as both try to assess attainment of project activities from information gathered. Each activity, however, refers to a different management function and time perspective.

Monitoring is primarily a tool to feed back information to project managers on what is actually happening on the ground with regard to the delivery of project inputs and their use to create outputs. It is carried out on a continuing basis so that prompt corrective action can be taken when necessary. Evaluation, on the other hand, is concerned with assessing, after the event, the level of success of the project (or part of the project), and whether this particular design of the project can achieve its longer development objectives.

Here, the tasks of the formulation team are to determine:

- Where in the project organization structure the M & E function is to be located, the composition of staffing, and to whom the group is responsible.

- The major items of information (key performance indicators such as inputs into the project, outputs, costs, revenues) to be collected by the M & E unit, how they should be presented, the type of reports to be produced by the unit and at what intervals, and for whom. At this point the team should also indicate the principal items of information to be collected in the baseline survey.

There may be a risk of bias if the M & E function is established within project management. For this reason governments and donor agencies often insist that it is institutionally located elsewhere, sometimes in a specialist unit. It is important, however, from the point of view of the team that monitoring is made to work effectively by ensuring that the information it collects (a) is of value to project management, (b) is passed to that management quickly, and (c) is used by it. Very often the most effective way to bring this about is to recognise monitoring as a project management function and place it within the project. Evaluation, on the other hand, will base much of its work on the monitoring information collected, and can be placed outside the project where it is and can be seen to be independent of the project.

Differences in Public/Private Sector Procedures (Step 10)

The activities of the two main tasks described in Step 10 are all relevant to public sector projects (Models B, C and D), particularly so when they are large projects with many components. Also, some of these activities are pertinent to private sector projects. Private sector projects (Model A) sponsors usually have a clear view at the outset of how the proposed project will fit into the overall company structure. As noted above, however, this issue can present substantial problems to a formulation team working in the public sector (Models C and D). The organization of project beneficiaries will in almost all cases be a matter of concern only in public sector projects. In contrast, issues related to project management are likely to be a common area of concern in projects in both sectors, although again it is likely that more attention will be necessary in those of the public sector.

Step 11 - Project Cost and Revenues Estimation and First Financing Proposals

Individual team members working on the design of individual project components (Step 9) will have, at the same time, costed their proposals as this is an important factor in selecting preferred design options. Together with estimation of costs for the selected project organization and management structures (Step 10), the final task for the team in Phase III is to consolidate the separate budgets into one cost estimation for the project as a whole, phasing the costs as appropriate, and preparing a proposal for the financing of the project (although in some cases the formulation team may have been instructed not to consider the question of financing at this stage). Thus the three main activities in Step 11 are:

- Completion of estimates of all individual components costs.

- Consolidation of component costs and of organization and management costs into overall project costs.

- Preparation of a financing plan to show how the project costs are to be met.

A number of rules are applicable with regard to what costs are considered and how they are calculated. They are described briefly here.

The cost of all project inputs, however small, have to be borne by someone. Inputs which are to be procured with cash are financial costs, and have to be taken into account in this Step. Inputs procured without cash, such as the use of family labour to dig a pond, etc., are non-financial costs. These are not directly the concern of the team in this

Step. Nevertheless, it is important that the team identifies and quantifies them for two reasons. First, the work to be done by agents may be a significant factor affecting the adoption rate of fish farming; second, such costs have to be taken into account later (Step 13a) when the economic analysis of the project is undertaken (public sector projects).

Any resource flow or input used by project activities and which would not be used in such activities if the project did not exist, is a project cost. In this category would fall, for example, existing field staff and facilities which are used for the benefit of the project. The costs of such staff and facilities are not usually included in the project budget, being regarded instead as in-kind contributions by the agency providing them. By the same token, all new staff recruited and new equipment purchased which would not otherwise be recruited or purchased through another source, are regarded as project costs; this is because these expenditures are truly additional to other expenditures and will be incurred only because of the project exists.

In public sector projects the estimated cost of contingencies, usually expressed as a percentage of a base cost, is added to the total at the end. In these projects, therefore, it is important in calculating unit costs to be as accurate as possible, and not add some margin for error for each item, otherwise the contingency accumulates and becomes a significant factor in the overall project cost. In private sector projects, where costing is done at a more detailed level than is general in the public sector, it is more usual for contingency elements to be calculated in respect of each item.

The information generated in this Step is used firstly to construct the financing plan for the project (Step 11d), and subsequently in the financial and economic analyses (Steps 12 and 13, respectively). It is necessary, therefore, to develop project revenue information, complementary to the cost information to complete these analyses.

In public sector projects, the financial and economic analyses are carried out on an incremental basis, that is the project is valued by reference to the difference between the with- and without-project situations. The period when the team is assembling the with-project costs and revenues will be, in many cases, an appropriate time to assemble the corresponding without-project information for use in the subsequent financial and economic analyses (Steps 12 and 13). The information assembled will be for both the project as a whole and for the principal agents.

Step 11a - Completion of estimation of individual component costs

Project costs for the purpose of Step 11 consist of all additional cash expenditures on goods and services which are required during the life of the project. It is necessary to distinguish between:

- Capital costs, mostly one-off expenses, incurred during the investment or development phase (such as land purchase, civil works, site development, machinery, equipment, breeding stocks, studies and engineering, and start-up technical assistance).

- Recurrent costs, normally calculated on an annual basis, incurred during the operational phase (such as seed, fertilizers, insecticides, feeds, fuels, lubricants, wages, salaries, repair and maintenance, replacement and maintenance purchases, supplies, utilities, rents, taxes, and insurances). Equally important for calculating recurrent costs are bioprogramming estimates of stock of fish on hand in any one month. This affects the logistics of feedstuffs (delivery, storage, and consumption) and cash-flow (Step 12). Here the contribution of an experienced farmer is important so that the logistics accurately reflect a farm situation.

After identifying the cost items as falling into either the capital or recurrent costs categories, the team should estimate the unit costs of each item, i.e. the civil engineer would estimate the cost of earth moving (US$/cubic meter), the mechanical engineer the purchase prices of pumps, and daily running costs, the fish farming expert would find the cost of feed delivered to the farm, and so on. Table 3 shows investment costs (before calculating contingencies, see below) for an aquaculture project in China. The Table shows the calculation of base costs from quantities and unit costs.

The cost information developed in this Step is, as noted above, used in the financing plan and the financial and economic analysis of the project. While the economic analysis is concerned with the costs and benefits to the economy generated by the project as a whole, it will be necessary for the construction of the financing plan and for the financial analysis, to have cost information for each of the principal agents. Clearly, these will differ according to the type of project. In a public sector production project (Model B), for example, costs would be required at the farm level, for processors, for the credit institution, etc. In many projects, farmers and other agents will differ considerably. In this situation, the team has to identify typical categories, noting the number of agents in each. Farmers, for example, may be categorized according to size of holding and project zone; fish traders may be categorized according to size and area of operation, etc.

The incremental costs of project components can then be calculated by multiplying the planned number of units to be used over and above what would exist without the project.

The physical inputs of the project are usually calculated over the life of the project (many externally-assisted public sector projects are assumed to have a life of about 15 years, but the period may vary according to the type and size of project). Cost estimates are usually made on a year-by-year basis over the life of the project. Table 4 shows the investment and operating costs of a shrimp farming project in China.

In separate tables, cost should also be broken down into:

- Materials which are imported into the country, with costs in foreign currencies.

- Materials normally obtained from domestic sources, with costs in national currency.

- Skilled personnel inputs with a component of foreign experts, and also foreign currencies.

- Unskilled labour inputs.

- Profits and trade margins.

Table 3

Phased investment costs of a prawn breeding sub-project in China
(¥RMB '000)

	Unit	Quantity 1989	Quantity 1990	Quantity 1991	Quantity 1992	Quantity 1993	Quantity Total	Unit Cost	Base Costs 1989	Base Costs 1990	Base Costs 1991	Base Costs 1992	Base Costs 1993	Base Costs Total
1. INVESTMENT COST														
A. Sea Dyke a/														
Earthwork	000 m3	100	68	-	-	-	168	2.68	267.0	181.6	-	-	-	448.6
Sub-Total Sea Dyke a/									267.0	181.6	-	-	-	448.6
B. Inlet and Outlet Canals														
Earthwork	000 m3	543	543	-	-	-	1,086	2.68	1,449.9	1,449.9	-	-	-	2,899.7
Main Sluice Gates	No	-	2	-	-	-	2	15.01	-	30.0	-	-	-	30.0
Sub-Total Inlet and Outlet Canals									1,449.9	1,479.9	-	-	-	2,929.7
C. Bridges and Sluice Gates														
Bridges	No	4	5	-	-	-	9	110.51	442.0	552.5	-	-	-	994.5
Sluice Gates	No	400	404	-	-	-	804	4.86	1,940.0	1,959.4	-	-	-	3,899.4
Culverts	No	-	1	-	-	-	1	361.15	-	361.2	-	-	-	361.2
Sub-Total Bridges and Sluice Gates									2,382.0	2,873.1	-	-	-	5,255.1
D. Pumping Station (60 m3/sec)														
Earthwork	000 m3	18	18	-	-	-	36	2.68	48.1	48.1	-	-	-	96.2
Stonework	m	1,000	1,500	-	-	-	2,500	0.13	125.0	187.5	-	-	-	312.5
Concrete	m3	230	230	-	-	-	460	0.16	34.5	34.5	-	-	-	69.0
Reinforced Concrete	m3	2,480	2,480	-	-	-	4,960	0.46	1,116.0	1,116.0	-	-	-	2,232.0
Axial Flow Pumps	No	4	4	4	-	-	12	100.01	400.0	400.0	400.0	-	-	1,200.0
Transformers	No	2	2	-	-	-	4	50.01	100.0	100.0	-	-	-	200.0
High Voltage Panels	No	11	11	-	-	-	22	13.61	149.6	149.6	-	-	-	299.2
Low Voltage Panels	No	10	10	-	-	-	20	5.01	50.0	50.0	-	-	-	100.0
Miscellaneous Equipment	LS	-	-	-	-	-	-		300.0	350.0	150.0	-	-	800.0
Installation	LS	-	-	-	-	-	-		100.0	200.0	100.0	-	-	400.0
Sub-Total Pumping Station (60 m3/sec)									2,423.2	2,635.7	650.0	-	-	5,708.9
E. Feeder Roads b/	KM	-	10	10	-	-	20	50.01	-	500.0	500.0	-	-	1,000.0
F. Prawn Pool														
Earthwork	000 M3	3,000	4,000	3,000	-	-	10,000	2.68	8,010.0	10,680.0	8,010.0	-	-	26,700.0
Sluice Gate Controls (2 ton)	No	150	200	50	-	-	400	0.51	75.0	100.0	25.0	-	-	200.0
Sluice Gate Control (30 ton)	No	150	200	50	-	-	400	0.61	90.0	120.0	30.0	-	-	240.0
Sluice Gate Control (50 ton)	No	5	6	5	-	-	16	1.01	5.0	6.0	5.0	-	-	16.0
Sluice Gate Control (100 ton)	No	2	3	-	-	-	5	1.61	3.2	4.8	-	-	-	8.0
Sampans	No	100	200	100	-	-	400	5.01	500.0	1,000.0	500.0	-	-	2,000.0
Miscellaneous Equipment	LS	-	-	-	-	-	-		1,250.0	1,250.0	1,250.0	-	-	3,750.0
Installation	LS	-	-	-	-	-	-		650.0	670.0	680.0	-	-	2,000.0
Sub-Total Prawn Pool									10,583.2	13,830.8	10,500.0	-	-	34,914.0
G. Power Plant														
Transformer Room	M2	-	165	-	-	-	165	0.6	-	98.2	-	-	-	98.2
Boiler and Storage	M2	-	98	-	-	-	98	0.6	-	58.4	-	-	-	58.4
Office and Dormitory	M2	-	72	-	-	-	72	0.6	-	42.9	-	-	-	42.9
Equipment		-	-	-	-	-	-		30.0	100.0	100.0	-	-	230.0
Electric Poles		-	-	-	-	-	-		15.0	16.0	-	-	-	31.0
Power Cables		-	-	-	-	-	-		172.0	200.0	200.0	-	-	572.0
Installation		-	-	-	-	-	-		170.0	200.0	200.0	-	-	570.0
Miscellaneous Equipment		-	-	-	-	-	-		18.6	18.6	18.6	-	-	55.8
Sub-Total Power Plant									405.6	734.0	518.6	-	-	1,658.2
Total INVESTMENT COSTS									17,510.8	22,234.8	12,168.6	-	-	51,914.1
Total									17,510.8	22,234.8	12,168.6	-	-	51,914.1

a/ Earthdyke 11 km long

b/ 20 km

Table 4

Investment and operating costs of a project in China for new shrimp ponds
(¥RMB million)

Years	1	2	3	4	5	6	7	8-19	20
Production Area (ha)[a/]	-	667	1 333	2 000	2 000	2 000	2 000	2 000	2 000
Costs									
Investment costs	24.50	39.50	27.20						
Operating costs									
Seedling[b/]	-	1.00	2.00	3.00	3.00	3.00	3.00	3.00	3.00
Feed - processed[c/]	-	2.75	5.50	8.25	8.25	8.25	8.25	8.25	8.25
- fresh[d/]	-	1.08	2.16	3.24	3.24	3.24	3.24	3.24	3.24
Fertilizer[e/]	-	0.02	0.03	0.05	0.05	0.05	0.05	0.05	0.05
Energy - electricity[f/]	-	1.29	2.58	3.87	3.87	3.87	3.87	3.87	3.87
- oil[g/]	-	0.02	0.05	0.07	0.07	0.07	0.07	0.07	0.07
Labour	-	0.89	1.78	2.66	2.66	2.66	2.66	2.66	2.66
Other costs[h/]	-	0.70	1.40	2.11	2.11	2.11	2.11	2.11	2.11
Working capital[i/]	3.00	3.00	3.00	-	-	-	-	-	(9.00)
Sub-total	3.00	10.75	18.50	23.25	23.25	23.25	23.25	23.25	14.25
Overhead costs[j/]	-	1.00	1.60	2.00	2.00	2.00	2.00	2.00	2.00
Total costs	27.50	51.25	47.30	25.25	25.25	25.25	25.25	25.25	16.25

e/ Construction of 200 ponds (each of 3.33 ha) in each of years 1-3
b/ 300 000 seedlings/ha, at ¥RMB 5/1 000
c/ 3.75 t/ha at ¥RMB 1 100/t
d/ 13.5 t/ha at ¥RMB 120/t
e/ 30 kg/ha at ¥RMB 800/t
f/ Total of 24 234 kw at ¥RMB 160/kw
g/ Total of 90 t at ¥RMB 800/t
h/ Maintenance, labour protection, compensation, etc.
i/ Represents short-term working capital to meet costs prior to harvest. Estimated on the basis of costs being spread over a 6-month operating period (May-October)
j/ Includes management charge paid to Breeding Company (2% of sales), insurance, other office overhead costs, etc.

- Taxes and duties (noting that imported equipment exempt from customs duties may not necessarily be tax free).

These cost breakdowns will facilitate the analysis (Step 13) when economic costs and benefits are calculated.

As already noted, the usual practice in the private sector is for the team to calculate the contingency elements in respect of each item. The team must remember that the safety margins which each cost estimator adds to his respective component can magnify the total cost. This is particularly important for capital costs of construction projects, such as hatcheries and farms. When production components are bioprogrammed from little factual data, biologists and technologists frequently add generous margins for error. This information is then transposed to engineering drawings, which are then costed with further margins for mechanical inefficiencies, safety, and back-up systems. For example, the biologist may have research information about the lethal and tolerable levels of oxygen for the species in question, but knowing the fluctuations due to the build up of ammonia, photoperiod changes, temperature changes, and altitude, etc., he specifies a required design level well within these levels. The design engineer translates the total oxygen requirement into water delivery. However, a mechanical delivery system of pumps and pipework has its own inefficiencies and resistances, and the engineer calculates these into his design, also adding a small margin of safety for extreme conditions. Thus, the final design of the hatchery or farm may have pumps and pipework considerably over-engineered for the task. Increases in costs for water delivery are exponential rather than linear, and consequently construction costs are greatly increased, as are subsequent operating costs of the pumps. These increases are further magnified in the contingency costs, which are a percentage of base costs. As a result, total costs may well be inflated.

It is important, therefore, that it is agreed, within a team working on private sector project formulation, where the safety margins are to be applied and by whom; team members who may have little experience in costing (for example, perhaps the aquaculture technical staff and biologists) should be aware of their limitations in this area.

Step 11b - Consolidation of component costs and organization and management costs into overall project costs

This activity is typically the responsibility of the financial analyst or economist in the formulation team, or possibly the team leader. The task involves the aggregation of the cost estimates for each component (production, support, social, and organization and management) into a consolidated table of costs for the project as a whole. Usually, except for small and simple projects, separate tables are prepared for capital and recurrent costs. All the projects costs have to be phased on an annual basis over the life of the project. This is largely a mechanical task. Care should be taken at this point to ensure that capital replacement costs are entered into the table at the appropriate points.

As noted above (Step 11a), the treatment of contingency costs differs according to whether the project is in the public or private sector. In the private sector contingencies are estimated on an item-by-item basis and, as a result, total project costs are not inflated. In the public sector, contingencies are added after all the other project costs have been aggregated and phased on an annual basis. Unless care is taken, this method will inflate total costs.

All project component costs (from Step 9) and project arrangement costs (Step 10) are aggregated and phased on an annual basis (phasing of overall project costs). When this is complete, a physical contingency, preferably not exceeding 5% of the base costs but never more than 10%, is added. Clearly, however, the degree of uncertainty associated with cost estimates depends to a large extent on the nature of project activities. For example, in a project to construct, equip, and operate a national aquaculture centre, with components of production, research, and training, all the elements are readily identified and can be costed with a high level of accuracy. In a project to improve inland fish production through small-scale farming, the potential number of participants cannot be estimated with any accuracy, and therefore the physical contingency has to be relatively large.

The team should estimate the costs of project components on the basis of present-day prices. The task of estimating how much additional finance will be required to meet price escalation in the cost of project components will be primarily the responsibility of the financial analyst or economist. It is likely, however, that the rate of cost inflation of different items will vary, and the economist/financial analyst should seek the advice of the appropriate technical specialists about price trends for the inputs with which they are most concerned. The inflation of the physical contingency costs, as well as the base costs, should not be overlooked.

Step 11c - Estimation of project revenues

This activity is confined to production projects, whether they be in the public or private sectors. It is usually undertaken by the economist but, in those teams where there is a marketing expert, he may undertake, or collaborate in the task.

To a large extent this task is a mirror image of Step 11a. First, the team has to make an estimate of revenues accruing to the relevant agents (for example, farmers, processors, credit institutions, etc.), grouped in the same categories as adopted in Step 11a, without the project. The information generated in Step 5b ("Projecting supply, demand and price changes") will be relevant to this part of the activity, but not until the project zones had been demarcated and specific agents selected will it have been possible to have undertaken this task with any accuracy. Second, revenues of the same agents, or categories of agents, with the project, are estimated.

In public sector projects where there are foreign currency earnings attributable to the project, the project revenues should be broken down into domestic and foreign currency earnings, with and without the project.

Step 11d - Preparation of a provisional financing plan

The annual financing requirement of the project is the amount by which the project costs (cash outflows) exceed project returns (cash inflows) each year. In investment projects the initial years normally show major cash deficits. The total financing requirement is therefore the sum of the annual cash deficits throughout the life of the project.

A project is financed from one or several of the following sources:

- Government budget allocations and/or loans.
- International aid and/or public loans.
- International and/or domestic commeroial loans.
- Equity investment.
- Proceeds from the sale of project outputs.

The usual practice of external assistance agencies is to contribute mainly to expenditures during the investment period, and not to expenditures during the operational phase, thus limiting their potential funding commitment in level and time. A further restriction on the type of lending by some agencies is a reluctance to finance capital expenditures in domestic currency, preferring to finance only foreign currency expenditures. It follows that the team should be aware of the policies and procedures of the financing agencies concerned, and to keep proposed contributions within such constraints.

Experience indicates that it is important not to assume that the government, particularly of a developing country, will be able to meet the balance of the project costs not financed by external agencies. For a production-oriented project, the formulation team has to establish the official policy on project cost recovery of both the project sponsor and the government, and to what extent they require the project to be self-financing. This may result in two questions which have to be answered by the team in relation to the specific design of the project:

- How much of the surplus, which accrues to project beneficiaries as a result of their participation in the scheme, can reasonably be taxed?

- Assuming that in principle there is scope for cost recovery, can a practical and effective method of tax collection be devised.

For production-oriented and other projects, which are to be partially funded through external assistance and where a government contribution is required for project implementation and its subsequent operation, the team has to make a judgement of whether, given the priority attached by the government to the project, the government's share of the costs will be forthcoming. When the team has any reservations about the matter, they should be raised as an outstanding issue in the project report, to be resolved directly between the government and the funding agency.

Table 5 shows an example of the sources and application of funds in a project. It gives the breakdown of project cost recovery between the project revenues and external assistance loan, and the net financing requirement for which a government contribution would be required.

Apart from directly financing part of the project costs through taxes, levies, and service charges, the beneficiaries may also finance investment costs indirectly through borrowing and repayment of long-term loans. These can also be included in the financial organization of the project. Careful analysis is required here to show, with a reasonable degree of accuracy, the income position of the beneficiaries. This is necessary to ensure that it is likely to be feasible to finance the required part of the project in this way, i.e. (a) that beneficiaries' net incomes will be sufficiently high to enable the repayments to be made and, (b) borrowing is attractive from the beneficiaries' viewpoint.

Table 5

Financing requirements of a new fish-pond sub-project in China
(¥RMB)

	1	2	3	4	5-7	8	9-10	11	12	13-19	20
REVENUES											
technical service fee	205	308	410	513	513	513	513	513	513	513	513
marketing fee	137	205	274	342	342	342	342	342	342	342	342
taxes	684	1026	1368	1710	1710	1710	1710	1710	1710	1710	1710
contract fee	6265	6265	6265	6265	6265	6265					
total revenues	7291	7804	8317	8830	8830	8830	2565	2565	2565	2565	2565
OPERATING COST											
administration(1.5%)	205	308	410	513	513	513	513	513	513	513	513
taxes	684	1026	1368	1710	1710	1710	1710	1710	1710	1710	1710
subtotal operating cost	889	1334	1778	2223	2223	2223	2223	2223	2223	2223	2223
operating incomes	6402	6470	6539	6607	6607	6607	342	342	342	342	342
INVESTMENT COST											
infrastructure	28515	-	-	-	-	-	-	-	-	-	-
vehicle	447	486	-	-	-	-	-	447	486	-	-49
tools	444	-	-	-	-	-	-	-	-	-	-
technical assistance	-	28	28	-	-	-	-	-	-	-	-
overseas training	10	77	77	-	-	-	-	-	-	-	-
local training	206	220	259	131	-	-	-	-	-	-	-
research	38	76	76	38	-	-	-	-	-	-	-
subtotal investment cost	29660	887	440	169	-	-	-	447	486	-	-49
net income before financing	-23258	5583	6099	6438	6607	6607	342	-105	-144	342	391
FINANCING											
loan received(70%)	20755	621	308	118	-	-	-	-	-	-	-
repayment(9.9%)	4212	4338	4400	4425	4425	250	-	-	-	-	-
net financing	16543	-3717	-4092	-4307	-4425	-250	-	-	-	-	-

In practice, the analysis of the financial position of the beneficiaries may fall within the next Step. However, it is wrong to think of project design, including the preparation of the financing plan and project analysis, being in separate compartments. Instead, they should be seen as complementary, the financial analysis showing where changes in the project design are needed.

Differences in Public/Private Sector Procedures (Step 11)

The three tasks of this step are broadly applicable to all projects, with the exception that contingencies (11b) are treated differently in the public sector (Models B, C, and D). The detail in the costing process in the formulation of projects in the private sector (Model A) is such that sound judgements can be made about the physical contingencies required in respect of each item. In public sector projects costing is not carried out to such a detailed level.

Illustrations of Phase III: Project Design (Steps 7-11)

Project 1 - A Commercial Fish Hatchery Project in a Mediterranean Country

Following preparation of the prefeasibility study report, which recommended a smaller production figure than originally desired and a nursery located at the lower level of the site, the team leader had returned to the country and presented the report to the investor.

After reviewing the report the investor accepted the recommendation to construct the nursery nearer sea level, but wanted the original production target to be maintained. Without need of further studies, he confirmed the design criteria (Step 8c) in his discussions with a local architect engineering design firm who then offered their proposal (Step 8d) to undertake the work (Step 9) in two stages, preliminary design and final design. On signing the two-part contract for time and materials, he gave the firm the prefeasibility report and bioprogramming document.

After three months the firm presented the preliminary design and revised costs which it estimated were now within about 15% of accuracy. The firm pointed out that the project would now carry increased costs due to the higher costs for blasting a larger area of the foreshore than was originally conceived for both the nursery and the pumping station. Furthermore, the costs of the hatchery would be increased because of the need for materials which would withstand corrosion from seawater, and the pumphouse on the shoreline needed substantial protection from wave exposure.

The cost of implementing the project was now becoming larger than the project sponsor had anticipated. Although it would delay the project by probably a year, he decided to seek an investment subsidy from the EEC which, for a hatchery, could be as high as 70% of construction costs. The architect and engineer firm was requested to complete the final design drawings, and the original consultants were rehired through another lump-sum contract to prepare two feasibility reports. These were for attachment to the proposals for financial assistance, first from the government of the country, as a necessary preliminary stage, before presenting the second to the EEC, as part of a national package of structural projects for aquaculture.

Final design documents were then used to solicit construction bids from three local contractors, which narrowed the accuracy of costs to 5%. Using this figure, the consultants prepared the feasibility reports which included the organization and management plan for the project (Step 10) and the financing plan (Step 11).

Project 2 - A Brackishwater Shrimp Farming Project in a Southeast Asian Country (Steps 7-11)

A project identification team had carried out the preliminary formulation and design (Steps 3-6), investigating the ranking potential sites where traditional fish ponds might be used for shrimp farming through rehabilitation of canals and deepening of ponds.

As part of its interim report, the team had recommended that a national institution should carry out a detailed topographic, soil and water quality survey prior to further engineering design work being undertaken in project preparation. It was decided by DOF that too much time would be lost seeking external assistance for this work but that it should be carried out immediately using national funds (Step 7). The surveys were completed in six months and a land suitability map produced for each of the proposed project areas as part of the report.

In the meanwhile, the interim report was approved by the donor and DOF, and funding for the preparation stage released. The project preparation team was mobilized immediately after the land and water survey had been completed. It consisted of the following disciplines: team leader, economist/financial analyst, shrimp-farming expert, civil engineer, mechanical engineer, institutions expert, and a credit expert.

The team started its work with the project objectives unchanged (Step 8a). Forming part of this Step but with the work being carried out by the economist in the field and when the report was being prepared, information was collected on farmgate prices for shrimp and milkfish, and an analysis was made of international market trends. The surveys carried out in Step 7 had provided the team with the basic resource information which, combined with other information available on such matters as location and capacities of processing plants, access roads, etc., allowed it to demarcate accurately the project zones in each of the two main regions. It was then a simple matter to select the producer agents who, it was hoped, would participate in the project. The zone demarcation and agent selection in this case confirmed the preliminary work and, similarly, the system and technology remained as previously identified (Steps 3c, 4c).

The team now had the information to have defined project zones in each of the two regions assuming in all 40 000 ha and containing 15 000 ha of ponds for about 4 000 farmers. The primary, secondary and tertiary canals associated with these zones were included in the demarcation (Step 8b).

The topographic soil-and water-quality surveys enabled the team engineer to estimate fairly closely the amount of work required on the ponds in each of the project zones (Step 9a). In this project, the work required on the production support components (particularly canal rehabilitation but also hatchery and the research centre construction), formed the major part of the project. With the project zones now demarcated and with the information from the survey, the engineers produced outline designs for the canal rehabilitation work, pumps, gates, etc., and working with the shrimp-farming expert they produced outline designs for the hatcheries and research centre. Concomittantly, they made satisfactory estimates of costs, using earthmoving costs provided by the Department of Hydraulics (DOH), building construction costs provided by DOF, and equipment costs obtained from suppliers (Step 11a).

Credit was also seen as an important support component of the project. A credit programme was therefore designed by the credit expert, based on information about the number of farmers and the experience of the agricultural development bank of providing credit to farmers in the agricultural sector in the project areas. The credit expert spent much of the duration of the mission on the design of a scheme acceptable to all the institutions involved (Step 9b), prior to costing it (Step 11a). At this stage, therefore, the assumption made during project identification (Step 4b) that credit was required, was translated into a project component.

With the amount of detailed information available, project scheduling (Step 9d) was considered not to present undue difficulties and the team leader, in consultation with the other team members, produced a simple bar chart (Step 9d).

The organizational choices were regarded as being relatively straightforward, as it was considered there was no reason why the project should not use the existing organizational structure whereby DOH would be responsible for canal rehabilitation, DOF would be responsible for the construction and operation of the research centre and hatcheries, and for farm activities including extension, and the Agricultural Credit Bank would be responsible for the disbursement of loans to farmers and their recovery (Step 10a). A small PMO was proposed through which all activities would be coordinated and project funds disbursed (Step 10b). Technical Assistance was included within the project for the operation of the hatcheries and research centre, mainly in the form of fellowships. No provision was made for monitoring and evaluation (Step 10c).

Using the individual component costs provided by the engineers and the institutions expert, the economist produced the required cost analyses (Steps 11a, 11b) and, with the price information (Step 7), estimated the revenues of the main agents, that is farms grouped by size in each project zone, the Agricultural Credit Bank, and the Government (in the form of tax receipts) (Step 11c). In earlier discussions between the external assistance agency and the Government, it had been agreed that although the agency did not normally finance expenditures in national currency, it would do for this project since almost all the expenditure was in national currency. A preliminary financing plan was then prepared. This showed not only the financing requirement of the project as a whole but also the income/expenditure positions of each type of farmer and of the agricultural credit bank (Step 11d). These were satisfactory.

This part of the project formulation process, excluding the topographical, soil and water survey, took four months to complete.

Project 3 - A Shellfish Farming Enhancement Project in Southeast Asia

This public sector regional project concerned increasing the production and upgrading the quality control of farming mussels and oysters in the five Southeast Asian countries, with external assistance from the EEC.

As the whole process of project formulation had been fully budgeted from the beginning, the team embarked directly on the project design phase from the conclusion of the previous phase. This phase was scheduled to take 10 months. Its terms of reference clearly specified that the team's main task was to select critically from among the different options and variants upon them which had been identified, to produce the best possible project within a total budget not exceeding US$ 20 million, with a 5-year implementation period.

Steps 7 and 8 presented no particular difficulties. In the design of the individual components (Step 9), however, where actual choices had to be made to keep or discard more than half of the possible components identified previously, there were intense discussions within the team. Information on some of its preliminary choices leaked to Government representatives, leading to severe criticism from the representatives of two countries in the project steering committee.

The formulation team leader dismissed the team member who appeared to be the source of the leak. This action severely disrupted the work of the team and, after a few weeks, the team leader also left. An acting team leader was selected from within the team who directed the team to analyse each option. Two months later the new team leader prepared a series of component analysis sheets, clearly showing the advantages and disadvantages of each option. These were presented to the steering committee members.

The discussion on the basis of the sheets brought a clear consensus. All production components were discarded, and only a training and extension production support component (Step 9b) retained. It was decided that the project should concentrate on the introduction into the area of a sanitary control network (both at the levels of production and of the harvesting/marketing process), the transfer of depuration plant technology, and the setting up of an export marketing organization.

From this point the team had no further difficulty in implementing Steps 10 and 11, and keeping the project within the budgetary limits established.

Project design was finally completed 11 months after it started, a remarkable result given the the problems which had arisen.

Discussion Topics

Points for discussion concerning the above illustrations might include the following:

- the extent to which the type of project being formulated has influenced the way in which the formulation process has been undertaken;

- action taken, or decisions made, which would seem likely to prejudice the formulation process and/or project implementation.

PHASE IV: Analysis of Expected Results

Purpose and Outputs

By the time this phase has been reached, the project has been designed and costed in detail. It is necessary now to make a detailed assessment of the effects and impact of the project once it is implemented. The purpose of Phase IV is to confirm that the tentative conclusions made in the outline specification of a possible project (Step 6) remain valid now that the project has been designed in greater detail.

The approach to the analysis of the results differs considerably between private sector projects and those in the public sector. In the former, analysis is usually confined to the financial analysis and, increasingly, the environmental analysis. The analysis of results in the public sector is usually wider ranging, and the measure of profitability is rarely expressed in the same terms as in the private sector but rather as IRR, net present value, and the benefit/cost ratio.

In the treatment of this phase, attention is directed primarily to the requirements on the project planner of project formulation in the public sector. In Step 12d, however, a brief outline is given of the approach to be adopted where the private sector is concerned.

Two main outputs which should result from this phase are:

(i) A series of assessments with regard to the project's likely impact in terms of:

- (a) the income position of the main agents involved and the profitability of their operations;
- (b) the efficiency with which the project uses resources which could be used elsewhere in the economy;
- (c) the resulting social changes; and
- (d) the environmental impact.

(ii) Conclusions regarding the extent to which, and manner in which, the project's objectives and rationale conform to the priorities and principles of overall national development policy.

The work in Phase IV has four steps; a financial analysis (Step 12), an economic analysis (Step 13), a social analysis (Step 14), and an environmental impact assessment (Step 15). The activities of these steps are illustrated in Figure 9, and described in detail in the following text.

Step 12 - Financial Analysis

Financial analysis treats the project as a business proposition and assesses whether or not it adds to the net worth of the business or, when a number of different types of agents are involved, how profitable participation in the project is likely to be for each main category of agent. To make this assessment, the team values the costs and benefits of participation by the agents (for example, farmers, processors, credit institutions, government departments, etc.).

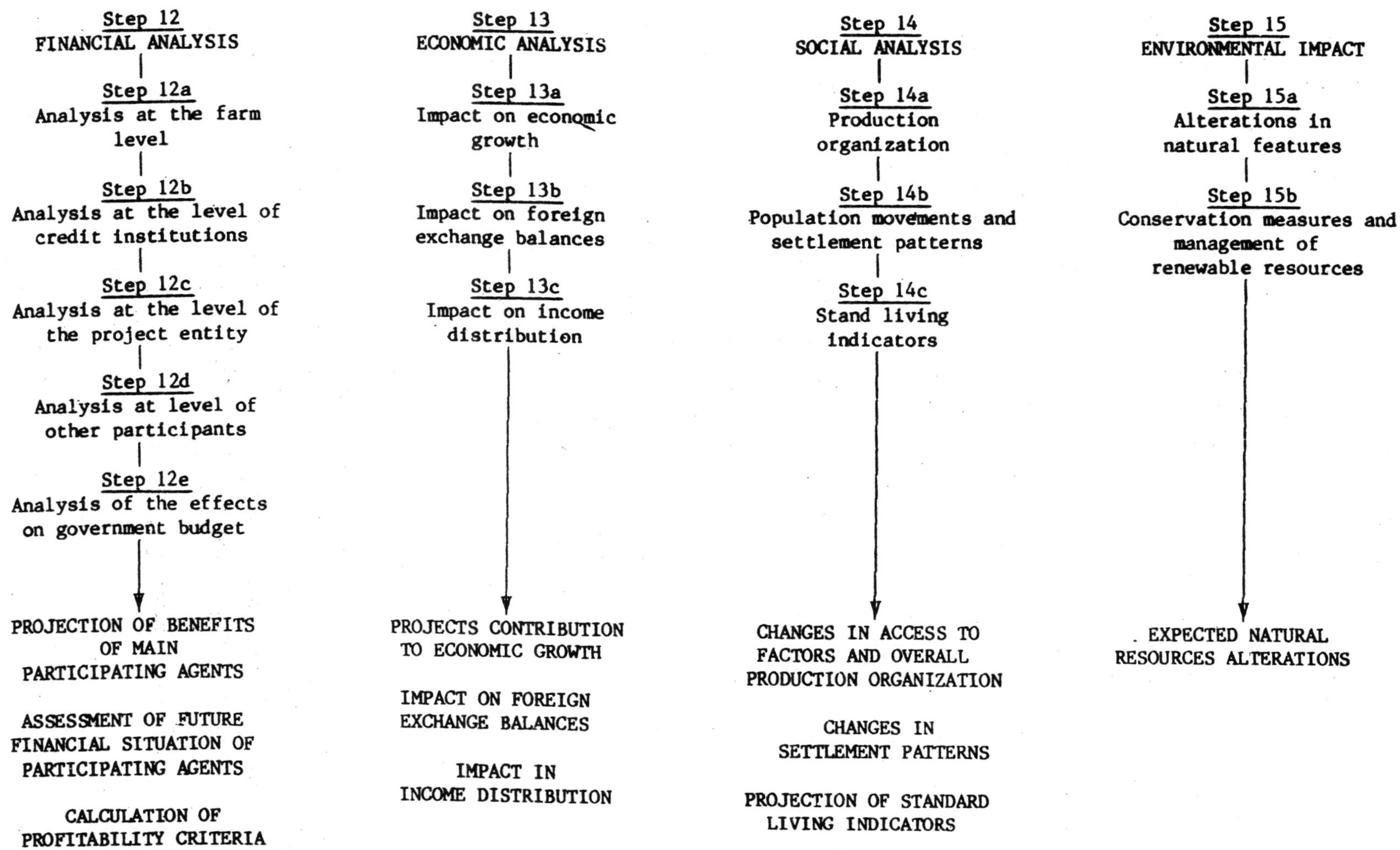

Figure 9 Phase IV. Analysis of expected results

Also, in addition to documenting the expected impact of the project on the liquidity, credit-worthiness, financial efficiency, and earnings, etc., of the various agents involved, the team should, as much as possible, use Step 12 to feed back the results of financial analysis into the selection of design criteria. (It is for this reason that the feasibility study of commercial projects in the private sector is often considered to be a task integrated into project design (Phase III), not one which is separate to and following upon that phase).

There are two main approaches in financial analysis, namely (i) cash flow and benefit/cost analysis, and (ii) income and expenditure analysis; the former is the more important analytical tool for project formulation.

Cash flow analysis traces movements of money in and out of the entity concerned (farm, hatchery, company, bank, etc.) when cash receipts and payments actually occur (in practice usually on an annual basis). Two stages are considered in the analysis, namely before financing and after financing. In the former, all credit transactions are excluded from the inflows and outflows. Its purpose is to estimate when and how much money the agents concerned will need to be able to invest as planned (perhaps in the case of farmers), or meet financial targets (perhaps in the case of financing institutions). In after financing analysis, all credit receipts (loans) and payments (debt service) are included. The purpose of cash flow analysis after financing is to determine whether the agents who have to borrow to finance investment will receive sufficient cash income to repay their loans, and retain a part of their increased earnings.

The benefit/cost analysis is similar to the cash flow analysis except that, in addition, it takes account of non-cash receipts and payments (products grown for direct consumption by the household, loans repaid in kind, etc.).

Income and expenditure analysis values inputs and outputs not as they are bought and sold but when they are physically consumed or produced. This takes into account any changes in assets between the beginning and the end of the accounting period. This type of analysis is an essential tool in assessing the viability of project proposals from an individual agent's perspective but does not convey a complete picture of the agent's economic situation; for example, a full assessment of the "income" position of (say) a farm should take into account the changes in assets between the beginning and end of the accounting period.

As noted below (Step 12d), complete financial analysis of enterprises often requires financial data to be presented in a number of different ways, including the balance sheet and sources and applications of funds statement. In general most analyses do not go into any detail in this area and what will almost certainly be required for most types of agents in most projects is cash flow and benefit/cost analysis.

The analyses referred to above are always made to show the expected impact of the project by comparing the "with-" and "without-project" situations, that is, an incremental analysis. It is computed as follows:

incremental net cash flow - equals net cash flow with project - less net cash flow without project

Alternatively, the same result can be obtained by computing:

incremental net cash flow equals incremental cash inflows less incremental cash outflows

The types of analysis referred to above are made under assumptions of certainty. In practice, of course, farm operations and indeed all commercial operations carry risks. To deal with risk, analysts usually use "sensitivity analysis". This means simply trying out plausible alternative values for the key variables in the appraisal and seeing what the effect is on the project as a whole. In aquaculture projects, variations in variables, such as the cost of feed, growth rates, survival, prices of product, etc., can be very substantial, so the analyst should take expert advice in selecting variations which are realistic. For example, the team might decide that the maximum the price of feed might increase over that price used in the analysis is 20%, that product prices might be 15% lower than used in the analysis and that survival rates might be 10% lower. These changes would be then used in the analysis to determine the worst possible scenario which is foreseen.

A spin-off from sensitivity analysis is "switching values". Each variable in the financial model can be altered at will by the analyst. The switching value is the value any variable (as above) would have to reach as a result of a change in any unfavourable direction before the project no longer meets the minimum level of acceptability, as indicated by one of the measures of project worth. For example, if the minimum internal rate of return is determined to be 12%, switching analysis allows the analyst to determine how high the price of feed can rise, or how low prices can fall before the rate of return falls below 12%. The particular value of switching value analysis is that it conveys the relative importance of different variables in terms of the sensitivity of project performance to them.

Step 12a - Analysis at the farm level

In aquaculture production projects the basic economic unit is usually the farm. For the financial analysis the inputs and outputs of the farm are expressed in the farm model. Although models reflect the actual situation as closely as possible, they necessarily have to simplify it. In a project, farms may be of different sizes, using different sizes of ponds, and sometimes there may be different technologies being employed. The project analyst will construct a farm model for each type of farming activity. The basic "building blocks" of farm models are activity budgets in which the inputs and outputs are expressed in unit cost terms. For example, production may be expressed in terms of revenue per hectare (or reference pond, cage, or enclosure) multiplied by the surface area or number of reference units in the model; the unit of seed may be 1000 pieces (or individual fry), and the total cost of seed will be the number of units multiplied by the cost per unit.

This picture of total inflows and outflows represents the farm budget before financing. It may need to be completed by showing any grants, subsidies, taxes, or levies involved. The analyst then adjusts the budget by including any loan to the farmer, and debt service he may incur; this gives the farm budget after financing.

When constructing farm budgets it is important for the team analyst to use prices paid by or to the farm household. These are obtained from field observation, or from parallel or illegal markets, which are used to confirm any official prices. Other points which the analyst should bear in mind are (i) the need to exclude household labour from the calculation, (ii) similarly, the need to exclude the cost of land from the calculation if the household owns it, and (iii) the need to include household consumption of farm production at market prices.

In making models and budgets it is advisable to avoid trying to be over-realistic, so making the model time consuming and complex. It is better to concentrate on capturing the key changes expected between the present, the implementation, and the fully operational stages of the project.

A farm budget should include:

- Before financing:

  Inflows (benefits): gross value of production (including home consumption and non-cash transactions, valued on a farm-gate price basis), receipts in kind, subsidies/grants, residual values (land excluded).

  Outflows (costs): investment expenses, cash operation expenses, payments in kind (labour supplied by the farm household itself is not counted as a cost), taxes/levies.

- After financing:

  The same Inflows augmented by loan receipts;

  The same Outflows augmented by debt service payments.

The difference between inflows (positive) and outflows (negative) is the household net benefit from farming, first calculated before financing, and later after financing.

In some cases some members of a farm household work partly or fully outside the farm. It is then necessary to calculate if implementation of the project has an effect on any involvements outside the household budget, including revenues of that activity.

Farm and household budgets are compared "with-" and "without-project" to calculate the incremental farm/household net benefit resulting from the project. This will reveal the attractiveness of the project to the farmers in financial terms.

Table 6 shows the calculation of net incremental benefit in a project in China to a farmer using fish-cum-pig culture. In this example a farming household is working on a contract basis. Here, the institution placing the contract bears the financing charges and consequently they do not have to be carried in the farm budget. In this example, the team has assumed that the price of fish falls from Y.8.0 to Y.3.5/kg, but the return on family labour increases from Y.4 per working day to Y.39 before falling back when pig production declines.

Some caution is due here. Where a farming activity is concerned with one particular activity, for example, fish-cum-pig culture, the adoption of

Table 6

Net incremental income per 1-ha modified pond with pig (contract household) in China
(¥RMB)

	1	2	3	4	5	6
REVENUES						
fish production(kg)	3000	4500	5625	6750	6750	6750
pig production(kg)	1425	2850	2850	2850	3250	1425
fish revenue	11400	17100	21375	25650	25650	25650
pig revenue	5700	11400	11400	11400	11800	5700
less without project fish revenue	6555	6555	6555	6555	6555	6555
total inc income	10545	21945	26220	30495	30895	24795
COSTS						
fingerlings	2250	-	-	-	-	-
fry	225	322	386	386	386	386
tools	-	-	-	-	-	80
feed	210	1190	2446	3938	3938	3938
maintenance and others	240	240	240	240	240	240
fertilizer	207	235	263	290	290	290
disease prevention	213	213	213	213	213	213
technical service fee	171	257	321	385	385	385
interest for fish working capital(9.9%)	374	-	-	-	-	-
marketing	114	171	214	257	257	257
tax	570	855	1069	1283	1283	1283
less without project production costs	3232	3232	3232	3232	3232	3232
total prod cost	1342	251	1919	3759	3759	3839
production cost for pig	6384	10054	10054	10054	10054	6384
interest for pig working capital(9.9%)	585	-	-	-	-	-
total incremental gross incomes	2234	11641	14248	16683	17083	14573
contract fee(20%)	1000	3330	3330	3330	3330	3330
net incremental incomes after contract fee	1234	8311	10918	13353	13753	11243
labour requirement(md)	340	340	340	340	340	340
return to family labour(y/md)	4	24	32	39	39	28

a simple farm budget (an incremental activity budget) which applies only to one hectare of the particular crop is suitable and convenient. In those farming activities where fish culture is one of many farm activities, it is preferable to develop a full farm budget, as in agricultural project analysis, which shows the combination of all farm activities. The failure to do so could result in harmful changes in resource allocation.

Naturally, production projects involve farmers putting in additional resources, of seed, fertilizer, labour, and often capital. Very often also farmers have to learn a new technology or apply a higher level of management expertise.

The non-financial factors which have to be taken into account by the team in their estimation of the rate of adoption of the project being introduced will have been incorporated if necessary into the social and technical investigations at the beginning of project preparation. At this point the team is concerned with the attractiveness of the financial inducements.

The team, therefore, has to make a judgement about whether the additional income which is generated for farmers by the project is sufficient to induce them to work harder to pay for the additional inputs which are required. Where the farmer has to make an investment, for a new pond or pump, his main concern will be whether the income generated will, in addition to enabling him to repay his loan and interest obligations, also pay for the additional inputs and generate a sufficient surplus to make worthwhile the investment risk.

Table 7 shows a financial analysis of a project in China for new fish ponds. In this case the ponds were to be constructed in an area previously unused and so no incremental analysis was necessary.

Among the means of measuring the value of investments are the benefit/cost ratio, the net present value, and the internal rate of return (IRR). These measures are dependent upon the discounting of the net cash flow. In the example in the Table, the IRR is 12%; no benefit/cost ratio or net present value is quoted.

Step 12b - Analysis at the level of credit institutions

The project-related cash flow of the credit institutions depends on the lending conditions proposed to the borrowers, as well as on the borrowing conditions borne by the institution itself for the resource it has mobilized or that it has been charged with managing.

The formulation team prepares year by year tables for each institution concerned, showing:

- Inflows (bank's own borrowing, repayment of principal from bank's borrowers, and payment of interest from bank's borrowers).

- Outflows (repayment of principal of bank's own borrowing, payment of interest on bank's own borrowing, loans to bank's borrowers, administrative costs).

The difference between inflows and outflows results in a yearly cash balance.

Table 7

Financial analysis of a project in China for new shrimp ponds
(¥RMB million)

Years	1	2	3	4	5	6	7	8-19	20
Production Area (ha)[a/]	-	667	1 333	2 000	2 000	2 000	2 000	2 000	2 000
Shrimp production (ton)[b/]	-	600	1 334	2 201	2 601	2 867	3 000	3 000	3 000
Gross revenue[c/]	-	9.0	20.01	33.02	39.02	43.01	45.00	45.00	45.00
Costs									
Investment costs	24.50	39.50	27.20						
Operating costs									
Seedling[d/]	-	1.00	2.00	3.00	3.00	3.00	3.00	3.00	3.00
Feed - processed[e/]	-	2.75	5.50	8.25	8.25	8.25	8.25	8.25	8.25
- fresh[f/]	-	1.08	2.16	3.24	3.24	3.24	3.24	3.24	3.24
Fertilizer[g/]	-	0.02	0.03	0.05	0.05	0.05	0.05	0.05	0.05
Energy - electricity[h/]	-	1.29	2.58	3.87	3.87	3.87	3.87	3.87	3.87
- oil[i/]	-	0.02	0.05	0.07	0.07	0.07	0.07	0.07	0.07
Labour	-	0.89	1.78	2.66	2.66	2.66	2.66	2.66	2.66
Other costs[j/]	-	0.70	1.40	2.11	2.11	2.11	2.11	2.11	2.11
Working capital[k/]	3.00	3.00	3.00	-	-	-	-	-	(9.00)
Sub-total	3.00	10.75	18.50	23.25	23.25	23.25	23.25	23.25	14.25
Overhead costs[l/]	-	1.00	1.60	2.00	2.00	2.00	2.00	2.00	2.00
Total costs	27.50	51.25	47.30	25.25	25.25	25.25	25.25	25.25	16.25
Taxes (5% of sales)	-	0.45	1.00	1.65	1.95	2.15	2.25	2.25	2.25
Net revenue	(27.50)	(42.70)	(28.29)	6.12	11.82	15.61	17.50	17.50	26.50

Financial Rate of Return = 12%

---

a/ Construction of 200 ponds (each of 3.33 ha) in each of years 1-3
b/ Yield assumption: year 1 - 0.9 t/ha; year 2 - 1.1 t/ha; year 3 - 1.3 t/ha; year 4 on - 1.5 t/ha
c/ Production valued (fresh, head on) at ¥RMB 15 000/t
d/ 300 000 seedlings/ha, at ¥RMB 5/1 000
e/ 3.75 t/ha at ¥RMB 1 100/t
f/ 13.5 t/ha at ¥RMB 120/t
g/ 30 kg/ha at ¥RMB 800/t
h/ Total of 24 234 kw at ¥RMB 160/kw
i/ Total of 90 t at ¥RMB 800/t
j/ Maintenance, labour protection, compensation, etc.
k/ Represents short-term working capital to meet costs prior to harvest. Estimated on the basis of costs being spread over a 6-month operating period (May-October)
l/ Includes management charge paid to Breeding Company (2% of sales), insurance, other office overhead costs, etc.

In this projection it may be necessary to test the sensitivity of the institution's viability to varying proportions of loans having to be written-off.

Step 12c - Analysis at the level of the project entity

If a specific entity, such as a project authority or company, is to be set up, projected financial accounts are prepared. The purpose of this is to check that it will be able to meet all its expenses (capital and recurrent costs) and to anticipate the amount and timing of its financial needs. The cash flows considered are only those received in or paid out by the project entity acting as an agent.

Step 12d - Analysis at the level of other project participants

Other project participants may include fish and shellfish traders, transporters, processing factories, etc., either individual agents or firms, existing or new ones.

The financial analysis for these participating agents follows the general rules for the computation of <u>incremental net benefit</u> (or cash flow), with a display for every future year of:

- operating inflows (sales) and outflows (cash operating expenditures, indirect taxes, increases in fixed assets) to obtain the "net benefit before financing";

- subsidies, added to the previous inflows - the new balance is also called "net benefit before (commercial) financing"; and

- financing inflows (long and short-term loans received, increases in liabilities and accounts payable, interest received) and the corresponding outflows (debt service, decreases in liabilities, and accounts receivable) - to calculate the "net benefit after (commercial) financing".

The net benefit before financing is a measure of the "intrinsic" financial viability of the agent and can be expressed as a return on all resources engaged by the agent (equity plus loan(s)), while the net benefit after financing indicates the return on the agent's equity alone.

Enterprise accountancy is a specialized and complex field. The purpose of presenting expected financial results is to assist decision-making about investment and to justify any particular financial treatment proposed (for example, tax exemptions). In project formulation the financial analyst should concentrate on producing incremental net benefit calculations together with selected financial criteria. After this, attention can be given to preparing a more comprehensive set of financial data (balance sheet, income statement, sources and uses of funds statement, etc.) if required.

Step 12e - Analysis of effects on the government's budget

The budgetary effects of a project are computed in the form of a <u>government cash flow account</u>. For each year of the project life, items which have to be estimated are:

Inflows:
- grant receipts from international aid agencies;
- borrowings from international or national sources (directly to the Government and not to national credit institutions);
- debt service receipts (paid to government by participatng agents);
- taxes (excise taxes, import duties, income taxes, related to project activities);
- user charges (if paid directly to a government agency);
- dividends (received from public firms participating in the project).

Outflows:
- debt service payments (capital and interest):
- grants, subsidies, and loans (conceded to participating agents);
- direct investment (infrastructure and service facilities);
- recurrent costs (direct project-related activities, handled by the Government);
- salaries paid to civil servants (involved or seconded to the project);
- other administrative costs (directly project-related).

The difference between inflows and outflows may show a deficit (negative cash-flow) or a surplus (positive cash-flow). A cost-recovery ratio is calculated on an incremental basis to measure the proportion of public outlays recovered from participating agents.

Differences in Public/Private Sector Procedures (Step 12)

As noted above the five tasks of Step 12 are all related to public sector projects (Models B, C, & D). In addition to the key task (Step 12d) for private sector projects (Model A), some private sector projects would include Step 12a; for example, where a company intends to produce aquaculture commodities through the formation of a "nucleus estate", and where it will be necessary to estimate the impact on farmers of different sets of financial conditions.

Step 13 - Economic Analysis

Economic analysis during Phase IV of project formulation is an appraisal of the project in terms of its contribution to the general objectives of economic development policy for the country as a whole.

Common objectives of national economic development policies formulated by most governments, particularly in developing countries, are:

- Sustained economic growth through the efficient use of scarce national resources.

- Steady progress towards, or maintenance of, a satisfactory balance on the country's foreign exchange account.

- Equitable income distribution among nationals.

- Balanced government budget.

The fourth objective has already been analysed (Step 12e). The other three are addressed below (Steps 13a, 13b, and 13c).

Step 13a - Impact on economic growth

Economic analysis is concerned with flows of "real" resources. These are either inputs, the use of which by the project produces a shortfall in their availability to other projects, or outputs, which add to the community's material well-being or to its stock of wealth.

There are two basic approaches to economic analysis in project formulation, namely the "shadow pricing" (or "accounting pricing") method and the "effects" method. The approach which is selected will depend mainly on the preference of the sponsoring and financing agencies.

The shadow pricing method has been considerably advocated by international development agencies. Its principle is to define the project costs and benefits in terms of their value to the society as a whole, by:

- Putting a value on the use of resources which, though not priced on any market, nevertheless have real production value, e.g. the labour a farm household devotes to working on its own farm.

- Adjusting values of project inputs and outputs with identifiable market prices, whenever these prices are considered not to reflect the true values to the economy of the items in question.

- Eliminating from the analysis all "transfer payments", i.e. payments made for which no goods or services are received in return (such as taxes, duties, subsidies, loans, debt service, social security, and social security charges).

Table 8 shows the economic analysis of the same shrimp pond project in China for the which the financial analysis was given in Table 7. In the economic analysis shadow prices have been used which have had the result of slightly increasing the IRR over that calculated in the financial analysis.

In the economic analysis the net present value of the investment cost at 12% is computed to be Y. 8.23 million, that is, the sum which is calculated to accrue as a result of the investment over and above that which the same sum would have earned if invested at the opprortunity cost of capital of 12% (the computed average rate of return on new investment elsewhere in the country).

The calculation of shadow prices is a specialized task, and a detailed description of the methods used would not be appropriate here. It is worth noting, however, that because this calculation is a complex task much time

Table 8

Economic analysis of a project in China for new shrimp ponds
(¥RMB million)

	Years 1	2	3	4	5	6	7	8-19	20
Economic Value of Production 1/	-	8.19	18.21	30.04	35.50	39.13	40.95	40.95	40.95
Economic Costs									
Investment Costs	22.60	36.20	25.60	-	-	-	-	-	-
Operating Costs									
Seedling 2/	-	0.91	1.82	2.73	2.73	2.73	2.73	2.73	2.73
Feed - Processed 3/	-	2.29	4.58	6.87	6.87	6.87	6.87	6.87	6.87
- Fresh 4/	-	1.18	2.35	3.53	3.53	3.53	3.53	3.53	3.53
Fertilizer 5/	-	0.02	0.04	0.06	0.06	0.06	0.06	0.06	0.06
Energy 6/	-	1.16	2.32	3.47	3.47	3.47	3.47	3.47	3.47
Labour 7/	-	0.51	1.01	1.52	1.52	1.52	1.52	1.52	1.52
Other Costs 8/	-	0.76	1.53	2.29	2.29	2.29	2.29	2.29	2.29
Working Capital	2.75	2.75	2.75	-	-	-	-	-	(8.25)
Sub-Total	2.75	9.58	16.40	20.47	20.47	20.47	20.47	20.47	12.22
Overhead Costs 9/	-	1.32	2.11	2.64	2.64	2.64	2.64	2.64	2.64
Total Costs	25.35	47.10	44.11	23.11	23.11	23.11	23.11	23.11	14.86
Net Benefits	(25.35)	(38.91)	(25.90)	6.93	12.39	16.02	17.84	17.84	26.09

Economic Rate of Return = 13.5%

NPV at 12% = Y8.23 million

1/ Production valued (fresh, headon) at Y13,650/ton.
2/ 300,000 seedlings/ha, at Y4.55 per 1000.
3/ 3.75 t/ha, at Y916 per ton.
4/ 13.5 t/ha, at Y131 per ton.
5/ 30 kg/ha, at Y985 per ton.
6/ Electricity conversion factor of 0.88
7/ Unskilled labour CF of 0.57.
8/ Non traded goods CF of 1.09.
9/ Non traded services CF of 1.33.

may be wasted calculating shadow prices for factors or items which are relatively insignificant to the analysis. It is better for the team to identify the critical values and concentrate on these. In aquaculture the most important shadow price is frequently that of labour. Very often, shadow prices for labour and other key variables, have already been computed by the national planning office.

The effects method has been developed in France and is used in several developing countries where French cooperation is most active. Economic growth is measured by increases in the national income, i.e. Gross Domestic Product (GDP), or Gross National Product (GNP). The value-added (VA) of an activity is the difference between the market price of the goods and services produced (P) and the cost of the intermediate consumption (IC) required to produce them.

Value-added measures the wealth created by the agent controlling the production process, and represents the value which the agent adds to the initial value of the inputs consumed in producing the new output. To calculate the value added by every participating agent, the cash-flow analysis tables are modified slightly into tables showing the yearly "operating accounts". In these tables:

- IC cumulates raw materials, other inputs, transportation, overheads, and other costs.

- VA cumulates wages, taxes, financing costs, and gross operating benefits (net operating benefit plus depreciation).

The aggregation of the value-added of all the agents directly or indirectly participating in the project represents the gross contribution of the project to:

- The GDP when the "domestic" value added (DVA) is considered, which makes no distinction between agents.

- The GNP when "national" value added (NVA) is considered, which is DVA minus wages, financing costs, and operating benefits paid to foreign parties (workers, institutions, and enterprises).

Thus, incremental value-added measures the effects of the project on economic growth in terms of additions to GDP or GNP.

Whichever method is used, the bottom line of the formulation team's economist's calculations concerning the expected impact of the project on economic growth is an estimate of the economic surpluses or deficits created each year as a result of investment in the project, i.e. project incremental net income benefit (shadow price method terminology), or project incremental net value-added (effects method terminology).

This output is then tested by a sensitivity analysis for all those factors about which there may be some uncertainty.

Step 13b - Impact on foreign exchange balances

Assessment of the project's impact on foreign exchange balances follows general lines of any cash flow analysis. Inflows are gains in foreign currency (increased exports and/or decreased imports), and outflows are losses in foreign currency (increased imports and/or decreased exports).

Prices used in the calculations are actual prices registered at the national border, usually CIF prices for imports and FOB prices for exports. The net foreign exchange flow is the balance of inflows minus outflows. The net incremental foreign exchange flow represents the actual impact of the project by comparing the "with-" and "without-project" situations.

Step 13c - Impact on income distribution

As a general rule aquaculture projects are rarely sufficiently large to justify an analysis of income distribution. Where it does appear to be appropriate, it will be noted from the above that the value added is the sum of incomes accruing to agents, namely:

- workers: salaries and wages,
- government: taxes,
- financial institutions: payment of interest,
- enterprises: operating profit.

Thus, the breakdown of incremental income distribution should encompass, in net and incremental forms:

- rural incomes, with possible distinction between different categories of individual aquaculturists;

- wages of employees in aquaculture enterprises;

- operating profits of aquaculture enterprises;

- profits of individuals and enterprises serving the aquaculture production sector (craftsmen, suppliers, manufacturers, etc.);

- credit institutions' net incomes;

- operating profits of public entities administering the project;

- overall net budgetary effects.

Differences in Public/Private Sector Procedures (Step 13)

As noted in the introduction to Phase IV, all the three tasks of Step 13 apply primarily to public sector projects (Models B, C, & D).

Step 14 - <u>Social Analysis</u>

In this step the formulation team attempts to assess changes in the lives of individuals and communities as a result of the project.

How the analysis will be made will depend on the kind of project and its circumstances. In public sector production projects (Model B) it is often convenient to consider three areas of change, namely production organization, population movements, and standard of living. The work for each requires a synthesis of the knowledge and experience of the team as a whole, although the sociologist will have the responsibility for this part of the report. Each of the possible areas of change is considered below as a separate activity.

Step 14a - Production organization

Three questions indicate the scope of this task:

- How will relationships between those who have tenure of land or have water rights be affected by the project?
- How will the project affect access by different groups in the community to other factors of production?
- What changes can be expected in the production system?

Access to land or water exploitation rights is usually a sensitive issue. The formulation team must take care that the project not only conforms with existing rules and traditions, but also avoids any risk of appearing to be biased with regard to the "appropriation" of rights from others. For example, the adoption of a new technology by entrepreneurs may lead to the increased profitability of aquatic farming which may in turn lead to either the eviction of tenant farmers or the buying out by the entrepreneurs of small farmers who have not invested, or could not, in the new technology.

Access to other factors of production raises several issues of importance for the team to resolve. For example:

- Which farmers may not be able to afford inputs planned in the project?
- What are the risks for farmers using the proposed credit system?
- Which farmers may or may not be prepared to increase their use of inputs, such as feeds, fertilizers, or hatchery-reared seed provided by the project?
- Which farmers will use subsidized inputs as recommended, or divert them to other uses?

While the more dynamic farmers will quickly grasp the opportunities offered by a project, it is important that the formulation team considers measures to provide all farmers with opportunities to participate, without favouring a particular group.

Changes in the production system often result from projects. It is important that the team assesses the consequences of any such changes to existing relationships between farmers, and with other rural agents, such as terrestrial farmers or fishermen. For example, changes in the organization of production may produce social changes, such as division of labour in the family production unit, or the acceleration of a trend away from traditional extended families into nucleus estates.

Step 14b - Population movements and settlement patterns

Possible changes in settlement patterns should be identified by the formulation team. These may include restrictions on the movement of nomadic fishermen, villge concentrations, and urban migration, etc. Two kinds of changes may occur, namely:

- Permanent redistribution of the population, when a project creates areas of high development potential (for example, the construction of networks of canals in coastal swamp areas providing opportunities for farming new ponds).

- Temporary migrations, resulting from seasonal labour shortages or demands due to cropping changes introduced by the project.

Step 14c - Standard of living indicators

The composition of goods and services which satisfy "basic needs" of any community vary from one area to another. Specific project objectives to help satisfy these needs will therefore differ case by case. However, the following indicators are (in most cases) considered by the formulation team:

- Income distribution (as identified in Step 13).

- Employment, in terms of person-years of work created by the project.

- Nutritional effects, in terms of expected levels of daily protein and calorie intake relative to present levels.

- Improvement in domestic water supplies.

- Project impact on health.

- Project impact on housing improvements.

- Project impact on quality of life of project participants (in general).

Differences in Public/Private Sector Procedures (Step 14)

As with Step 13, all three tasks of Step 14 apply mostly only to public sector projects (Models B, C, & D).

Step 15 - Environmental Impact

Rural and coastal aquaculture development projects invariably modify biological and physical processes which shape the natural environment. The interaction between project activities and the environment at times may be sufficiently complex or extensive to merit thorough analysis.

The environmental analysis aims at:

- Identifying renewable and non-renewable resources used by the project, and determining the ecological consequences of their use.

- Identifying sensitive points in local ecological systems which may be adversely affected by the project.

- Assessing the risk of pollution from aquaculture production and related processing activities.

- Assessing, in general terms, the extent to which major production systems designed for the project are ecologically sustainable in the long term.

Many countries have developed procedures for providing both environmental impact assessments and/or impact statements. In the former, the investor is usually required to make a general statement of the effects of the project on the environment; in the latter the statement is much more rigorous and may require all facts about the project and a detailed estimate of its effects on the environment. These efforts will have been identified by the team in Step 3b, general evaluation of the project area/sub-sector situation.

Step 15a - Alterations in natural features

For an aquaculture project, potential implications are analysed from two points of view:

- Possible alterations of the project environment due to other projects, particularly new agriculture or industrial projects, which may affect the project in the near term and long term.

- Possible alterations to the project environment due to the project itself, which may affect not only project activities, but also fishing, tourism, or other economic activities in the area.

So far as possible these effects should be quantified and incorporated within the economic analysis. It is not unusual for formulation teams to follow the approach outlined here with regard to effects on the project; less frequently teams incorporate into the economic analysis quantified estimates of the effect of the project (project externalities).

Step 15b - Conservative measures and management of renewable resources

The formulation team should investigate the likely impact of the project on any renewable resources which are used, such as trash fish for feeds or animal manures for fertilizers, which could be used in other activities, possible adverse effects on local biodiversity and genetic resources, introduction of parasites and/or diseases, etc. A "resource balance sheet" is then constructed which shows the expected impact of the project on such resources.

Differences in Public/Private Sector Procedures (Step 15)

All three tasks of Step 15 are necessary for all types of projects (Models A, D, C and D).

Illustrations of Phase IV: (Steps 12-15)

Project 1 - A Commercial Fish Hatchery Project in a Mediterranean Country

On learning that the estimated costs were higher than anticipated, the investor had decided to seek additional financing through an investment subsidy from the EEC. He had contracted a local architect/engineering design firm to prepare final design drawings and solicit construction bids, and then rehired the original technical consultants to prepare two feasibility reports, one for the national government and the other for submission to the EEC.

As appendixes to the feasibility reports, the investor requested the consultants to include a financial analysis of the project (Step 12a), and prepare cash flow tables and income expenditure tables. Finally, because of the increased area on the foreshore to be blasted, he requested the consultants to expand the environmental assesment (Step 15a).

The additional work in Phase IV by the consultants added a further six months to the project schedule.

Project 2 - A Brackishwater Shrimp Farming Project in a Southeast Asian Country

Financial analysis at the production agent level and for the agricultural credit bank had been carried out as part of Step 11. In Step 12, an incremental financial analysis was undertaken for each project area, and the project as a whole. The opportunity cost of capital had been determined by the national planning office to be 12%. With the exception of one project area, the FIRRs exceeded this level. When switching values at 12% were applied, the project was found to be most sensitive to adverse changes in the prices of feed and of product. However, the team did not consider it likely that the prices of these variables would change sufficiently to bring the FIRR below the opportunity cost of capital.

The EIRRs were found to be rather higher than the FIRRs, mainly because the shadow price of labour had been judged to be 50% higher than the financial price and transfer payments of farmers, notably levies for the use of water, were excluded (Step 13a). The impact on the foreign exchange earnings (Step 13b), was shown to be beneficial, as was the effect on incomes (Step 13c).

The social impact of the project (Step 14) was considered to be negligible. The environmental impact of the project (Step 15) was considered, with the social analysis, as an adjunct to the financial and economic analyses. This phasing, however, had adverse consequences so far as the time required for formulation was concerned. The possible scouring action of the water had been considered during design, and it had been concluded that it would have minimal adverse consequences. Similarly, screen forebays had been included at the design stage to disperse the physical effect of water intake. It became clear, however, when the environmental impact was being considered, that the disposal of spoil from the deepening and widening of canals and the deepening of ponds had been severely underestimated in the design stage. Much of the spoil would have to be dumped on the foreshore. This involved additional work on project design and further analysis to take into account the additional cost. The environmental effects of the dumping, however, were considered to be beneficial for future mangrove growth.

This phase was carried out within the same time frame as the previous phase. Thus no additional time was required.

Project 3 - A Shellfish Farming Enhancement Project in Southeast Asia

This regional project in the public sector was concerned with increasing the production and upgrading the quality control of farmed mussels and oysters in the five Southeast Asian countries, with external assistance from the EEC. During project design all the new production components were discarded, and the focus changed to training and extension in support of existing production, organization of a sanitary control network, and the setting up of a marketing organization.

Some difficulties and complexities arose in the analysis of expected results (Steps 12-15).

In the financial analysis it was found that the costs of establishing a network of water quality controls for the main production areas, and improving the sanitary quality of the production from a number of them, were high in relation to the financial benefits which could be expected, and so the FIRR was very small. In contrast, the economic (Step 13), social (Step 14) and environmental impact benefits (Step 15), were considered to be particularly high.

After discussions between the consultants and Steering Committee, the project was divided into a number of separate projects, some to be funded through grant assistance, including those which had the objective of improving water quality controls and sanitary quality, and the others through loans. Financial, economic, social and environmental analyses were carried out for each project.

The discussions and additional analysis, which were required, extended the schedule for this phase by five months, to ten months in all.

Discussion Topics

Points for discussion concerning the above illustrations might include the following:

- The role which project analysis might play in bringing about changes in project design; the implications of this for the arrangement and scheduling of project preparation.
- Similarities and differences in the approaches to project analysis and the techniques adopted according to the type of project.

PHASE V: Project Documentation and Submission

Purpose and Outputs

The purpose of the last phase of project preparation, Phase V, is to prepare the project report and submit it to the project sponsor.

After preparing and submitting the project report, formulation teams working on projects in the public sector will have generally completed their task. It is not usual for them to be involved in the subsequent administrative and financial negotiations relating to the project. In the private sector, however, the formulation team members, or their associates, may also be involved in the negotiation of the contract which is based on the project report, and subsequently manage and/or carry out project implementation.

The main outputs resulting from Phase V are:

- Physical preparation of a report (or set of reports) which documents all aspects of the team's work and provides a detailed project plan.

- Formal submission of the report(s) to the project sponsor accompanied, if requested, by special technical presentations to, or briefings of, the steering committee by the formulation team.

The work of Phase V is contained within one step (16), the activities of which are summarized in Figure 10 and described in detail in the following text.

Step 16
PREPARATION
OF THE PROJECT REPORT
AND SUBMISSION

Step 16a
The report structure

Step 16b
The contents
of the report

Step 16c
Report submission
and presentation

DOCUMENT OUTLINE CREATED

APPENDIXES PREPARED
(TEAM MEMBERS)

MAIN REPORT PREPARED
(TEAM LEADER)

EXECUTIVE SUMMARY PREPARED
(TEAM LEADER)

REPORT PRESENTED, FINALIZED
AND OFFICIALLY ACCEPTED
BY PROJECT SPONSOR

Figure 10 Phase V. Project documentation and submission

Step 16 - Preparation and Submission of the Project Report

Even the best field work and most careful and sophisticated analysis by the formulation team are all largely wasted if the findings cannot be properly conveyed to the project sponsor. Poor organization of material in a report is unfortunately often a major reason for the final breakdown in communication between the two. This first part of the Step is concerned with organizing the task of report writing and the production of the document. There are four main activities. They are:

- Deciding the structure of the report, both in terms of general organization (division of the main report itself and appendixes) and of detailed organization (internal organization of each document).

- Writing and production of the report.

- Submission of the report and its presentation.

- Finalizing the report.

Step 16a - The report structure

In most cases the team leader will be familiar with the requirements of the project sponsor regarding preferences for the structure of the report and how the sponsor's requirements fit this particular project. The sooner any project report format is known and communicated to the team members the better, as it is easier to have all provisional documents prepared in the proper format from the beginning rather than to reorganize them at this time. It will be recollected that at Step 8d it was suggested that the team leader discussed with the team the structure and contents of the report, and provided them with examples of the type of work expected.

An excellent way for judging what should be the length of the different parts of the report, their scope, and the detail required, is to identify who is going to read all the report, or only certain parts of it. Typical sections and potential readers are:

- The Summary. This may be targetted at top policy-makers, such as a Minister or Chief Executive, when the project formulation report is sent with covering memorandum from perhaps a Permanent Secretary or development division director, and all interested parties below.

- Main Report. This might be for non-technical senior managers, permanent secretaries in government ministries, and all interested parties below.

- Appendixes. These are usually read with considerable thoroughness by heads of technical departments, and technical specialists.

- Working Papers. These are for detailed examination by the appraisal team and, when required, by technical departments within the project sponsor. Both target audiences will use them to check the detailed assumptions behind the analyses.

The summary provides a self-contained picture of the project and all its implications, understandable without further reference, and in no more than two or three pages.

The main report is written in a consistent style, with well-balanced sections. It is an executive summary of the project based on material presented in appendixes, carefully edited to present a concise description of the objectives, findings, conclusions, and recommendations of the study.

Each appendix is a self-contained analysis of a single major aspect of the proposed project.

Each working paper is essentially a file on basic data, assumptions, estimates, and calculations on which alternative designs for project components in the respective appendix are analysed and subsequently based.

The contents of the main report are organized in sections arranged in sequence, and for a production project usually include the following:

- Summary.
- Introduction.
- Background.
- The project area or the sub-sector.
- The proposed project.
- Organization and management.
- Markets and prices.
- Financial and/or economic implications.
- Project justification and risks.
- Outstanding issues and follow-up action required.

Project formulation is a lengthy undertaking. It is easy to underestimate the time involved producing the report, and all the additional elements of proofreading, correcting, printing, binding, and circulation. Thus the team leader needs to give careful attention to scheduling completion of all working papers and appendixes, as some appendixes are dependent on others, and most appendixes have to be completed before the drafting of the main report can begin.

Step 16b - The contents of the report

Writing a good report is a skilled job. The team leader should follow five recognized basic principles writing reports, namely keeping the report:

- Simple, avoiding any unnecessary jargon.

- Direct, avoiding digressions, unless they serve a specific purpose.

- Lucid, being understandable by any intelligent reader and free of any ambiguity.

- Vigorous, sustaining the reader's interest, presenting arguments with convictions, and conclusions with confidence.

- Brief, but not to the point of sacrificing important material for the sake of limiting the length of a section of text.

While the structure of the report for an institution-building project (Model D) will be similar to that of a production-oriented project in the private sector (Model A), the contents will differ substantially. In the description of report writing below, attention is directed towards production-oriented projects.

The appendixes of a project formulation report are the building blocks from which the main report is constructed. The order in which they are presented in the report follows the logical sequence of formulation itself.

Each appendix is written in the form of a small monograph on a particular aspect or component of the project concerned, with adequate introductory and concluding chapters. Each can contain specialized bibliographies and should be correctly cross-referenced to other appendixes.

In the main report the purpose of the introductory section is to provide the context and setting within which the investment proposals have been formulated.

The background section introduces the important features of current policy on aquacultural development, reviews recent trends in relevant sectors, and describes the organization and functions of the major institutions concerned.

The section on the project area or sub-sector prepares the ground for explaining the project design decisions taken during formulation.

The rationale for the project includes a concise description of all major features of the project, presents a summary cost estimate and financing plan, and explains recommended procurement, accounting, and auditing procedures.

The organization and management section explains how various institutions and agencies will participate in implementing the project, and operating it subsequently.

The markets and prices section, where appropriate, shows how the inputs and outputs of the project will be traded, and indicates what is expected to happen to critical prices and price structures in the future.

The section on financial and/or economic implications describes the results of the financial and economic analyses.

Other impacts and main project risks are presented in the project justification section. This is often the first part of the main report to be read by the major decision-makers. Hence it is important to concentrate on particular benefits of the project rather than list results which may be expected.

Finally, the last section should draw attention to any major outstanding issues which must be resolved before project implementation can proceed. Thus it also contains suggestions of the steps to be taken to hasten progress.

The summary does not attempt to condense all the information contained in the main report. Its purpose is to highlight key findings and conclusions arising from the formulation study, and it gives the essential options upon which the investment authorities base their decisions.

Normally the summary is not longer than two or three pages in length. It is a good approach for the author to put himself in the place of senior managers and administrators who have to deal with a stream of information and briefing papers.

Step 16c - Report submission and presentation

Project sponsors might require that the report prepared at the end of Phase V be first submitted first as a draft final report, sometimes accompanied by a verbal presentation. Adjustments may then have to be made by the team before the report is accepted as the definitive project document.

A covering letter from the team leader, or organization responsible for project formulation, is a useful addition to a project formulation report when it is submitted, particularly if the report is mailed. This letter reminds the project sponsor of the circumstances in which the formulation study was commissioned, who carried out the report, and when. Any important revisions to the original terms of reference are noted. The letter concludes by recommending whether the project should be implemented, or not.

If a verbal presentation is requested, the presentation is made by the team leader, possibly assisted by one or two key team members. This presentation must also be carefully prepared. It should be concise, but complete in relevant detail, and well illustrated by several (about 10-12) visual aids. These are prepared professionally in the form of slides or transparencies. It is important that the presentation is self-sustaining, as often few of those attending have had the time to read the report fully, or at all.

Differences in Public/Private Sector Procedures (Step 16)

The three tasks of this step are applicable to all projects (Models A, B, C, & D).

Illustrations of Phase V (Step 16)

Project 1 - **A Commercial Fish Hatchery Project in a Mediterranean Country**

The high costs of the project had compelled the investor to seek additional financing through an investment subsidy from the EEC and he had hired the consultant to produce two feasibility reports.

The technical consultants prepared relatively short but detailed reports in English at their home office, complete with several appendixes, which included the technical programme, the financial analysis of the operations and an environmental assessment. One was prepared in the format for submission to the national government for a subsidy for which the report, financial analysis and impact assessment were translated. The second report was prepared in the format for submission by the government in its national package of requests to the EEC for structural projects, seeking an investment subsidy of 70% of construction costs, and remained in English.

The senior partner of the consulting company returned to the country with the documents, and submitted them to the investor (Step 16).

Project 2 - A Brackishwater Shrimp Farming Project in a Southeast Asian Country

The report was drafted and presented, together with the technical appendixes, in the format described in the text above (Step 16b). Working papers which had been prepared during the report writing were retained by the consultants in a form in which they were comprehensive and convenient sources of reference to an appraisal team. At the conclusion of the drafting of the report, the team was disbanded.

Project 3 - A Shellfish Farming Enhancement Project in Southeast Asia

No specific instructions were given by the EEC to the team with regard to the creation of the document outline and writing the report (Step 16), except for the general instruction of preparing it to international aid agencies' standards. The project, by that time, had attracted the interest of the European Development Fund (EDF), the World Bank (WB), the Asian Development Bank (ADB), and the United Nations Develoment Programme (UNDP).

When the report was submitted (Step 16c), it was immediately accepted and the formulation team was disbanded.

Discussion Topics

Points for discussion concerning the above illustrations might include the following:

- Differences in the procedures between the commercial project and those funded through official external assistance.

- The extent to which the reports were responsive to the needs of the respective situations.

4. PROJECT APPRAISAL

The third and final stage of project formulation is project appraisal. It only has one Phase (VI), concerned with project negotiation.

PHASE VI: Project Negotiation

Purpose and Outputs

Although the work of the project formulation team may in some cases be finished with Phase V, the project itself will not yet have been accepted for implementation. The transition of the project from completion of formulation to implementation can be negotiated in a number of ways because of different practices between the public and private sectors, and also within the two.

Differences result from differing attention given to the four tasks of project negotiation, and in particular the second, project appraisal. In a number of donor agencies, project appraisal takes the form of a thorough technical review of the entire project as presented by the formulation team. In other agencies, however, and also in the private sector, project appraisal is rarely carried out, and the sponsor negotiates implementation of the project on the merit of the project as presented.

Project negotiation may or may not require the presence of members of the project formulation team. In the public sector it is generally the case that project negotiation including appraisal is undertaken by the principals concerned; for example, between the government of a developing country and a donor agency. In the private sector, the project sponsor may or may not seek further help from the formulation team to represent the sponsor and/or describe the merits of the project to a financial institution.

The purpose of Phase VI is to negotiate the contract for implementation.

The main outputs of this phase are:

- Acceptance of the project by the project sponsor, with or without appraisal.

- Negotiation of the project between the project sponsor and financial institutions (in the private sector), and with the Government concerned and any implementing agency (in the public sector).

Step 17 - Project Negotiation

As noted above, in the public sector the procedures and preferences of the project sponsors differ. Very often, however, the project report is used as a basis for discussion between the project sponsor and other parties which may be involved (preliminary negotiation). The same report or, alternatively, an "appraisal report" prepared by the project sponsor, is used for discussion and agreement by the project sponsor internally (internal negotiation). The final negotiation of the project between the project sponsor and the government and/or implementing agency then follows.

In the private sector, after the project is accepted by the project sponsor it is taken to the financial institution for approval; or, when companies are financing the investment directly from their own resources, the project report will usually be first approved by the technical division concerned before submission to a higher authority for approval.

Project negotiation is often a complex and time consuming process, requiring much patience, understanding, and professionalism. A complete description of all the issues which may arise in negotiation is outside the scope of this document. Negotiation is considered here as only one step (17), with four principal tasks. This is illustrated in Figure 11.

Step 17a - Preliminary negotiation

When the project formulation report of a public sector project is accepted by the project sponsor, it provides the basis for discussions between the sponsor and any other parties concerned, to agree the terms of their participation.

These discussions may result in the need for some changes in the project. Although, for example, the formulated project may be satisfactory (in the sense that it fits both the terms of reference and the requirements of the area/sub-sector), changes in the policy of the project sponsor or government concerned may have occurred during the formulation process which would make some adjustments necessary. In some cases these adjustments may be extensive.

In the private sector, where few projects involve third parties, this step is usually omitted.

Step 17b - Project appraisal

Some project sponsors use the project formulation document throughout negotiation. Most development banks, and some large corporations, however, use a different approach. After analysing the project formulation report and making a decision favourable in principle to project implementation, these institutions prepare internally a project appraisal report. This is

Step 17
PROJECT NEGOTIATION

Step 17a
Preliminary negotiation

Step 17b
Project appraisal

Step 17c
Fulfillment of conditions by the government and/or implementing agencies

Step 17d
Project agreement

PROJECT APPRAISED AND ACCEPTED IN PRINCIPLE BY PARTIES INVOLVED

PROJECT NEGOTIATED AND CONTRACTED

Figure 11 Phase VI. Project negotiation

then presented to the institution's Board of Directors or other decision-making committee, for final approval.

The aims of appraisal are to:

- evaluate the financial , economic, and social objectives of the project;

- verify the procedures of the project formulation team;

- recommend the conditions which will ensure that the project objectives are met; and

- ensure that the proposed grant loan expenditure is in accordance with the policy of the financing institution.

Project appraisal is normally a process of verification of the situation in the field and a scrutiny of the report documentation. While the form of appraisal will vary according to the type of project, for production-oriented projects it will normally include the following aspects:

- technical, for example, engineering design and environmental matters;

- financial, for example, requirements for funds, the financial situation of the implementing agency, and of project beneficiaries when appropriate;

- commercial, for example, procurement and marketing arrangements;

- social, for example, sociological factors and expected impact of the project on certain groups (such as ethnic minorities and women);

- institutional, for example, organization and management arrangements, the requirement for arrangements of technical assistance, project monitoring and evaluation;

- economic, for example, project costs to the national economy, and the size and distribution of benefits.

The format and content of any appraisal report is specific to the agency, bank, or corporation concerned, and is their internal reference document. Thus its reflects their views, and not those of the project formulation team. It is on the basis of this report that the project is formally approved. It is also the reference document to which the implementation agreement is linked, and on the basis of which any subsequent evaluation is made.

Step 17c - Fulfillment of conditions by the government and/or implementing agency(ies)

This is usually a step only relevant to projects funded through external assistance. The decision of the board of the project sponsor is generally conditional. It is now time for those conditions to be met by the government and/or executing agency(ies). Such conditions involve:

- financial aspects (guarantees);

- ownership aspects (land rights, water rights);

- administrative aspects (delegation of responsibilities for project components or, if the project requires establishment of a new structure, effective establishment of that structure); and

- human aspects (assignment of staff, recruitment of new staff).

Governments may be able to comply with these conditions very quickly, but in other cases a period of several months (or even longer) may elapse before this step is completed.

Step 17d - Project agreement

In the public sector, when all the conditions outlined in Step 17c have been met, the agreement can be finalized. It can take different forms, according to the national legislative requirements, and the regulations of the agencies concerned. In general, however, an agreement usually states:

- the specific source of funds, to whom they will be released, and how;

- procedures to be applied in key contracts to be negotiated for implementation (such as engineering, construction, procurement of equipment, recruitment of consultants for technical assistance, etc.);

- the expected implementation schedule, critical steps to be noted, how, and by whom; and

- procedures of project supervision and monitoring, and responsibilities for these tasks.

The signature of the project implementation agreement marks the end of project negotiation (Phase VI), and therefore the end of project formulation as a whole. Project implementation can now begin.

In the private sector, agreement by the Board or financing institution will follow upon Step 17b.

Differences in Public/Private Sector Procedures (Step 17)

Negotiation is a part of every project (Models A, B, C, & D). However, the four individual tasks of negotiation are more typical of public sector projects (Models B, C, & D), particularly those which are relatively large and costly. Small private sector projects (Model A) require only two tasks (17a and 17d) to complete negotiation, while larger projects may require the three tasks (17a, 17b, and 17d).

Illustrations of Phase VI (Step 17)

Project I - **A Commercial Fish Hatchery Project in a Mediterranean Country**

The high cost of the project had compelled the investor to seek additional financing through an investment subsidy from the EEC. The investor had accepted the technical reports of the consultants which were required for the request of the financial assistance.

The investor, having already notified the national authorities of his intention to apply for the eec assistance, submitted both reports he had commissioned (see the illustration to Step 16), with a covering letter. The government department concerned, after some discussion with the investor, accepted them for appraisal (Step 17b). Three months later, the investor was notified that his project would receive a grant from the national government in the event of the EEC also providing financial assistance and that his project would be included within a number of aquaculture projects to be submitted to the EEC.

The government submitted its proposal to the EEC two months later. It contained an overall appraisal of the seven projects which it was submitting, together with its appraisal of each project attached to that project. The EEC then made its own appraisal (Step 17b), based on the respective project document and the government's appraisal. Nine months later, the investor was notified by the government that his project had been accepted by the EEC with three others out of the seven which had been submitted; as the project included a hatchery, it was eligible for a grant amounting to 70% of the construction costs.

A number of forms were then completed by the investor (Step 17c), and two months later the first payment was made to the bank. Negotiations were finally completed on receipt of this payment.

The investor then signed a contract with the local architect/engineer firm which had designed the hatchery for construction supervision services. A local building contractor was selected for the work, and ground clearance commenced two months later.

The period of negotiation, until implementation began, had taken about 18 months.

Project 2 - A Brackishwater Shrimp Farming Project in a Southeast Asian Country

The project report had been accepted from the consultants by the financing institution concerned (Step 16c). No changes were considered necessary as a result of developments which might have taken place during project formulation (Step 17a), and the financing institution proceeded to mobilize a team to carry out the appraisal (Step 17b). The team consisted of a team leader and an economist/financial analyst (both staff members of the financing institution), a civil engineer and a shrimp-farming expert.

The appraisal team made a two-week visit to the country, having discussions with the government departments concerned and with the agricultural credit bank. It then visited a selected number of project areas before returning to the capital for further technical discussions and to resolve the two "Outstanding Issues" raised in its report by the formulation team.

In other respects, the appraisal team found the report of the formulation team to be acceptable and, on its basis, the appraisal report was written which, thereafter, was the reference document for the project.

The government was able to satisfy the loan conditions with little delay (Step 17c), and the Project Agreement was signed nine months after the completion of project formulation.

Project 3 - A Shellfish Farming Enhancement Project in Southeast Asia

The project report of the formulation team had been accepted and the team disbanded (Step 16). A number of donors had already expressed interest in the project.

The report was circulated among all those donors who had expressed an interest in it. In the meanwhile, a Middle East based donor agency requested a copy. It then expressed a wish to use the project to provide assistance from the Gulf States to the two Muslim countries in the region.

Project discussion (Step 17a), now revolved around two possible projects, one for Indonesia and Malaysia only, with financing from the new potential donor and technical assistance provided through a consultancy company, and the other, for the region as a whole, to be financed through a trust fund established by the European Development Fund (of EEC) and UNDP, together with a loan provided jointly by the European Development Bank (EDB) and the Asian Development Bank (AsDB) for the construction of the depuration plants.

Finally, after further complex negotiations at different levels, it was decided that three project documents should be prepared, each project having its individual emphasis to fulfill the conditions of the donor agencies and the respective national governments (Step 17c). One project was thus prepared, signed and implementation begun, 15 months after delivery of the original project document; another began eight months after than, and the third six months later.

The three projects were:

- a three-year project for shellfish depuration plants in Indonesia and Malaysia, co-financed by the Middle East based donor agency and the governments of both countries;

- a regional five-year project for enhancement of shellfish culture targetted at improving the living conditions of small-scale fishermen, co-financed by UNDP and EDF, and implemented by FAO;

- a regional project to improve coastal water quality, co-financed by EDF and another bilateral donor (which was a last minute newcomer to the project), and implemented by UNEP.

Discussion Topics

Points for discussion concerning the above illustrations might concern the following:

- the use of the appraisal process in the three illustrations;
- the role of negotiations.

FURTHER READING

GENERAL

Baum, W.C. and S.M. Talbot, 1985. Investing in Development: Lessons of World Bank Experience, Oxford University Press, Oxford and New York

___________, ZOPP (an introduction to the method), Deutsche Gesellschaft fur Technische Zusammenbeit (GTZ) GmbH, Frankfurt am Main

___________, 1985. Preparing Agricultural Investment Projects, FAO Investment Centre Technical Paper No. 1, FAO, Rome

___________, 1986. Guide for Training in the Formulation of Agricultural and Rural Investment Projects, FAO, Rome

PROJECT MANAGEMENT AND IMPLEMENTATION

Bainbridge, J. and S. Sapire, 1974. Health Project Management: A Manual of Procedures for Formulating and Implementing Health Projects, World Health Organization, Geneva

___________, Agricultural Project Implementation, US Department of Agriculture and US Agency for International Development.

PROJECT PLANNING

Mulvaney, J., 1978. Analysis Bar Charting: A Simplified Critical Path Analysis Technique, EDI, The World Bank, Washington, D.C.

RAPID RURAL APPRAISAL

Gow, D., Rapid Rural Appraisal: Social Science as Investigative Journalism. In Finsterbusch, Kurt, Jay Ingersoll and Lyn Llewellyn, eds., Fitting Projects: Methods for Social Analysis for Projects in Developing Countries, Lynne Rienner Publishers, Boulder, Colorado

Kumar, K., 1987. Rapid Low-Cost Data Collection Methods for A.I.D., USAID Program Design and Evaluation Methodology Report No. 10, US Agency for International Develoment, Washington, D.C.

Odell, Malcolm, Marcia Odell and S. Franzel, 1985. Informal Survey Methods for Farming Systems Research, in Human Organization, 44(3):215-18

___________, 1986. Diagnosis in Farming Systems Research and Extension, Volumes I and II, for Farming Systems Support Program, University of Florida, Gainesville

DESIGN

Brown, M.L., 1979. Farm Budgets: from Farm Income Analysis to Agricultural Project Analysis, Johns Hopkins University Press, Baltimore and London

Brown, C.M. and C.E. Nash, 1988. Planning an Aquaculture Facility, Aquaculture Development and Coordination Programme (ADCP/REP/87/24), FAO, Rome

___________, 1989. The Design of Agricultural Projects: lessons from experience, FAO Investment Centre Paper No. 6, FAO, Rome